William L. Baily

Trees, plants, and flowers

Where and How They Grow

William L. Baily

Trees, plants, and flowers
Where and How They Grow

ISBN/EAN: 9783743402416

Manufactured in Europe, USA, Canada, Australia, Japa

Cover: Foto ©berggeist007 / pixelio.de

Manufactured and distributed by brebook publishing software (www.brebook.com)

William L. Baily

Trees, plants, and flowers

Trees, Plants, and Flowers,

WHERE AND HOW THEY GROW.

A FAMILIAR HISTORY OF

THE VEGETABLE KINGDOM.

BY THE AUTHOR OF
"OUR OWN BIRDS."

WITH SEVENTY-THREE ENGRAVINGS.

PHILADELPHIA:
J. B. LIPPINCOTT & CO.
1876

Entered, according to Act of Congress, in the year 1869, by

J. B. LIPPINCOTT & CO.,

In the Clerk's Office of the District Court of the United States for the Eastern District of Pennsylvania.

PREFACE.

There are few recreative pleasures to which the minds of children can be turned with more real advantage, than to the close observance of the beauties of the creation. We are everywhere surrounded by objects of deep interest, which attract but little attention because their appearance is so familiar. But it certainly is an evidence that we do not sufficiently appreciate the blessings which our Heavenly Father has conferred upon us, when we look with indifference upon the beautiful adornments of that world which his own hand hath created, and whose every charm he hath so evidently designed for our enjoyment.

How elevating are the thoughts which are begotten by the contemplation of these objects!

how soft and gentle are the influences which take possession of the mind, when it turns the channel of its musings from the bright world around it, to the great Fountain and Source and Creator of all.

There is probably no time in life when a fondness for the study of Nature may be more safely cultivated than in early childhood. It is then that the mind turns with true simplicity from the visible objects of its admiration, to the adoration of the Invisible who created them; and what more happy moment than this to instil into the mind of a child the great truths of that religion, whose sublime doctrines are so abundantly illustrated in Holy Writ by direct references to objects in the outward world? Our Saviour, in his teachings to his disciples, frequently alludes to the beauties of his Father's creation: "Consider the lilies of the field how they grow; they toil not, neither do they spin; and yet I say unto you that Solomon in all his glory was not arrayed like one of these."

The object of these pages is not so much to attempt a description of rare and curious plants, as, by presenting some of the leading principles of Botany in a familiar manner, to beget in the reader an interest in a study which will at once prove a profitable pleasure and a wholesome recreation. The flora of our own land possesses attractions to all classes, inasmuch as it lies within the reach of all. The schoolboy may gather, as he goes, gems from the grassy roadside, which, if they be but the simplest of flowers, will bear the impress of a Divine finger, and convey a lesson of deep instruction. Let us not then undervalue the least of His works, who created nothing in vain, but who,

> With consummate skill devised the plan
> That creation's every voice should whisper
> Words of peace, and joy, and hope to man

INTRODUCTION.

There is no time in the day better suited to the study of Nature than the early hours of the morning. Every thing is bright and fresh and beautiful; the sweet little songsters are warbling their sprightliest notes; the lambs are skipping merrily over the soft green sod, from which the morning sun has not dispelled the sparkling dew; the tall trees wave their heads majestically to and fro; the babbling brook murmurs its softest melodies; while upon the gently blowing gale is borne the rich fragrance of many a fresh opening flower.

The whole creation is exhilarated by the rest of the night; and the mind, as well as the body, being refreshed by repose, becomes more capable of enjoyment; and if we can go forth with our hearts laden with gratitude to our Maker for his blessings, we shall be able to see in all his works, the evidences of a superior wisdom, so adapting one part to another, that each seems to minister to the wants and the happiness of the other.

There are many objects of great beauty and interest, which can only be observed to advantage in the morning There are thousands of beautiful birds, which are active and busy in hopping about the trees, and sometimes about the very doors of our houses, early in the day, that retire to the deep shade of the forests, and pass the sultry hours of noon in the tops of the tallest trees, and there remain so quiet that few are aware of their existence.

Most flowers also open their petals to catch the first glimpses of the morning sun, at which time their colors are most brilliant, and their fragrance the sweetest. If, then, you would study Nature to advantage, you must devote to it the earliest hours of the morning.

Various Grasses.

CHAPTER I.

GRASSES—THEIR PECULIAR FORM AND APPEARANCE—MANNER OF GROWTH—ADAPTATION TO THEIR USES—WHEAT—BARLEY—MAIZE—OATS—RICE—PAPYRUS—PAMPA GRASS.

THE first visible objects that were created in the beginning, after the waters were gathered together, and the dry land had appeared, were grasses.

"And God said, Let the earth bring forth grass, the herb yielding seed, and the fruit-tree yielding fruit after his kind, whose seed is in itself upon the earth; and it was so. And the earth brought forth grass, and herb yielding seed after his kind, and the tree yielding fruit, whose seed was in itself after his kind; and it was so; and God saw that it was good."

The term grass, as it is sometimes used in Scripture, evidently includes a great variety of plants, as in the passage commencing with, "If God so clothe the grass of the field which to-day is and to-morrow is cast into the oven," &c., it undoubtedly alludes to the various smaller herbs which commonly grow in the fields, as in those days the stems of plants were often gathered by the poorer classes, and used for fuel.

Of what is commonly called grass, there are many varieties, some of them possessing properties which render them very useful; while others seem to be created for the special purpose of beautifying the earth. Children may often be seen plucking their tall straight stems, and seem to take much delight in arranging them into bunches; and we might suppose that the interest generally felt in the soft verdure and the cool and refreshing appearance of a luxuriant grassy field or lawn, would excite in all but the most indifferent a desire of knowing something more than that certain plants are called grasses, as an acquaintance with their structure and properties cannot fail to afford both instruction and entertainment.

The vegetable kingdom is divided into three great natural orders, called Acrogenous, Exogenous, and Endogenous, terms referring to their different modes of growing. To the last of these belong the grasses. The word Endogenous means ingrowing; that is, the increase in the growth takes place upon the interior of the stem, which is often hollow, though mostly filled with a soft pithy substance, which becomes harder as it nears the outer surface of the stem. The peculiar formation of the leaves of endogenous plants is also striking; the veins all run parallel to each other, mostly throughout their entire length, instead of branching off and forming the beautiful and prominent net-work so noticeable in others. To this order also belong a variety of plants differing widely from the grasses, such as some species of the Lilies, the Orchids, and many more, some of which will be mentioned hereafter.

The stems of most plants are much branched, but the formation of the grasses is peculiar, the stalk being mostly tubular and jointed, and quite simple, except where, in some instances, it is parted to give place to a cluster of flowers. The leaves are very long and narrow, and the flowers are variously arranged, sometimes scattered loosely upon the stem, as in the oats, sometimes in a short compact head, suspended from the ends of long and slender branchlets, as in the Rattlesnake's grass; and sometimes they are densely crowded at the end of the stem, in a lengthened spike, as in Wheat, Rye, and Timothy. Each flower is composed of concave valves, placed one over the other; they are most conspicuous when the beautiful white, yellow, purple or scarlet anthers, which are hung gracefully upon their slender filaments, project from the lips of the corolla: the plant is then in bloom, and who does not admire a specimen of fresh-blown Fox Grass or Timothy, especially when on some foggy morning the minute particles of moisture have settled upon the delicate stamens, giving them the appearance of being thickly studded with jewels?

We have said that many of these plants have jointed stems; this is observed in those whose leaves grow one above the other; each joint here answers the double purpose of giving strength to the stem and support to the leaf. But, in other varieties, the only leaves produced are what are called radical, or leaves growing from the root; with these the stems are not jointed, but receive additional strength from their being sometimes triangular or square, and mostly very fleshy and stout.

It is a remarkable provision of Nature, that those plants which appear to have been designed for food, either bear seed in great abundance, or are supplied with some separate provision for reproducing themselves; this is particularly noticeable in the grasses. Wheat, Rye, Corn, Oats, Rice, and Barley, which constitute staple articles of food, all produce their seed in great quantities; while in many species whose seed supply the wants of the birds, the roots are perennial and creeping, and are continually sending up suckers, thus increasing themselves many-fold by a distinct method.

Those seeds which require to be sown every year are reserved for the use of man, whose superior intellect teaches him the proper mode of rendering them useful; while the inferior animals depend entirely upon that provision which Nature has made for their supply.

How beautifully do these facts illustrate those words of our Saviour: "Consider the ravens, for they neither sow nor reap, which neither have storehouse nor barn, and God feedeth them!"

The cereal grasses, or corn plants, are very rapid in their growth, and in a surprisingly short time send up a tall hollow stem, divided by joints, where the leaves are inserted, one at each joint, on the alternate sides of the stem; each leaf embracing the stem like a sheath.

Wheat stands at the head of the cereal grasses for its great productiveness and utility. There are several cultivated varieties of Wheat, the origin of all of which is involved in obscurity. No where has it been found in a wild state; at least, botanists have not been able to identify any of the wild species with those in cultivation. There is no doubt of its great antiquity, grains having frequently been found enclosed with the mummies of Egypt. Some of these having been sown, have produced plants similar to those now grown in the Levant.

There is much to admire in a simple grain of wheat. It contains within itself a marvellous vegetative power, which, after having lain for centuries in the darkness and obscurity of an Egyptian tomb, is capable, if rightly managed, of being made to yield not only a new plant, with its abundant spike, but also plant after plant, and spike after spike, until the produce of this single grain might at length feed a family, the dwellers in a village, the inhabitants of a city, and even of a nation, or of the world.

Wheat, Barley, and Rice.

Barley, although not so much grown in this country as some other corn plants, is nevertheless the principal crop in some parts of the world. In Egypt and Syria it forms the staple grain for making bread. It is not capable of producing the beautiful white and fine-grained loaves that are made from wheat, as it lacks the glutinous properties which are necessary to facilitate the action of the leaven. It is therefore made into thin cakes without the use of yeast, and hence it has been called unleavened bread.

We read of such cakes in very early times. Some, most probably of Barley, are mentioned when the Lord appeared to Abraham in the plains of Mamre, and the patriarch said to Sarah, "Make ready quickly three measures of fine meal; knead it, and make cakes on the hearth." Gen. xviii. 6. The Arabs of the desert act precisely in this way now, when they entertain strangers, using Barley meal to prepare their hearth-cakes. And the bread used by our Lord when he fed the multitude with five loaves and two small fishes, is expressly stated to have been made of barley. John vi. 9.

It is much to be regretted that so great a quantity of Barley should be wasted in producing intoxicating liquors, so destructive to the peace of mankind, this being the best grain for distillation; and from the ready market thus obtained, the farmer is often led to grow barley, and neglect crops which would be useful in supplying the means of human subsistence.

The most important grain next to wheat is Maize, or Indian Corn, which is a native of America, and was cultivated by the Indians previous to the landing of Columbus. It may be a matter of surprise to some, that this plant should be ranked among the grasses; but a little examination into its structure and habits will at once show its connection with them. It attains a much larger growth than any other of the corn plants, although there are many other grasses which even exceed it in height.

Maize is extensively cultivated in every part of the continent of North America. In the western states it is particularly productive, as it sometimes yields, under the combined influence of the rich soil and genial climate, at the rate of eight hundred for one.

Maize has never been cultivated in Europe with success, the climate not being favorable to its growth; hence it is rarely seen in England, except where a few stalks are raised as curiosities.

Oats is probably the next grain of importance, especially in America, where it forms one of the princi-

Panicle of Oats.

pal crops, being extensively used as feed for horses and cattle; it is much easier of culture than wheat, and can be grown on soil that would scarcely produce a good crop of any other grain. In Ireland it is raised in great quantities, and together with potatoes, forms a considerable part of the food of the peasantry. Almost any climate is adapted to the growth of Oats. Good crops have been seen growing close to the line of perpetual snow, at the Glacier de Boissons, on Mont Blanc; and it is said to have been found in a wild state on the island of Juan Fernandes, in the Southern Ocean; but the few plants discovered there may have been produced by grains accidentally scattered by some of the pirates who infested those seas soon after the discovery of the island.

Rice is a native of warm climates, and differs in the mode of its cultivation from any other grain that is grown. Those spots where various animal and vegetable substances are washed down by rivers, are most favorable to its growth. The marshy parts of Hindostan and Carolina are among the chief portions of the globe where rice is brought to perfection. But the American rice is generally considered as being much better than that which is grown in the East Indies.

The rice-fields of Carolina lie adjacent to the larger rivers which flow toward the sea, and down whose rapid currents the floods of each Spring bring a fresh deposit of soil. They are enclosed in some places by neat embankments, through openings in which the water is allowed to run at such times as it is needed. The rice-seed is sown in a rich plot of ground, and allowed to attain the height of a few inches, when the plants are removed into the fields where they are to grow; the ground having been previously prepared by being overflowed with water until it is thoroughly saturated. These plantations require to be kept constantly moist, and as they usually lie below the level of the river, by opening the sluices in the embankments they are readily watered; this operation is repeated several times during its growth.

A field of young rice is a beautiful and interesting sight, but the great amount of decayed vegetation which the soil contains, renders the atmosphere very unhealthy, and few persons beside the negroes employed in cultivation can remain in the neighborhood with safety.

In the list of useful grasses we must not forget the Bulrush spoken of in the Bible. This appears to be no other than the paper-reed of the Egyptians. The ark in which the infant Moses lay among the sedges of the Nile, was made of this plant. Isaiah speaks of the paper-reeds by the brooks, (Is. xix. 7,) which undoubtedly alludes to the same, as it was found in great abundance, not only in the shallow parts of the Nile, but in the little streams in the vicinity.

Papyrus.

The Papyrus, or Paper-Reed, has a thick triangular stem, eight or ten feet in height, and is said always to turn one of its angles toward the current, as though to break the force of the waves. It formerly was very abundant in all parts of Egypt, Abyssinia, and Syria, but modern travellers describe it as now being very rare.

From the very earliest ages of Egypt, papyrus appears to have been used for various purposes, but especially for the manufacture of paper. Herodotus mentions paper made from it as being an article of commerce long before his time; he calls it *byblos*. This name, it is supposed, is the origin of the Greek word *biblion*, or book, whence comes the term Bible. The paper made by the ancients was formed of the pellicle found between the bark and the fleshy part of the stem; the pieces of this were united together until they were of a suitable size, when they were pressed and dried in the sun. Many manuscripts, written upon this paper, have been found in the swathings of mummies, which were perfectly legible, and are interesting on account of their great antiquity. Paper was made from the papyrus until the eleventh or twelfth century, when it was superseded by that made from cotton. The papyrus had also many other uses among the inhabitants of Africa. Boats of a considerable size were made of it, and are spoken of in the Scriptures. The tassel-like flowers which surmounted its tall straight stems were worn as coronals by illustrious men. The Abyssinians chewed the root and the woody parts of the stem, its sweet juice resembling liquorice. The stems, as well as being used for fuel, were also made into cordage, and woven into a coarse matting which was used for various purposes.

Those grasses, which seem to be created rather for the purpose of increasing our happiness by affording a pleasing and grateful prospect to the eye, than to minister to our comfort by supplying the wants of the body, are so numerous and so widely distributed, that all are familiar with some of them, and as any attempt to describe them would be useless in so small a compass as could be assigned them here, we shall only cite a single example, leaving it to the readers to enter more fully into the subject as their interest or pleasure may incline them, there being few, perhaps, who have not the opportunity of seeing them in profusion, as they exist almost everywhere, and

"Clothe all climes in beauty."

GRASSES.

Pampas Grass.

The Pampas Grass is a native of Brazil, and covers large tracts of country in the vicinity of Buenos Ayres, known as the Pampas, whence its name is derived. It grows to the height of twelve or fourteen feet. Many beautiful specimens are to be seen cultivated in the gardens of England, where the mildness of the climate is favorable to its growth. The annexed cut is a sketch of a plant growing in the grounds of

Stoke Park, which was long the seat of the Penn family of Pennsylvania celebrity.

These plants show to much better advantage when grown separately, as the long leaves, of which there is a great profusion, hang in thick tufts on every side. From the centre of these, the tall straight stems rise several feet above the mass of foliage, and are crowned with large plume-like heads of silvery-white flowers. Some of these separate plants have attained the height of fourteen feet, with a diameter of about eighteen feet; and occasionally they have been seen with as many as fifty heads of flowers.

How beautifully does this majestic species compare with some of the humble little varieties which are scattered over our meadows! and yet, while God hath given extraordinary grace and beauty to one, he has also endowed the others with qualities which render them none the less curious, and far more useful. How wonderfully are they adapted to the various uses assigned them! If animals were allowed to feed upon the foliage of the Pampas Grass, its beauty would be marred, and the life of the plant endangered; but not so with the meadow-grass; the more its leaves are cropped, the wider spreads the plant; the more it is trampled upon, the thicker and softer it grows; and so far from being killed by the frosts of winter, it seems only to gather more life from repose, and upon the return of spring it again shoots forth with renewed freshness and vigor.

CHAPTER II.

FLOWERS — THEIR VARIOUS FORMS AND COLORS — PARTS OF A FLOWER — ARRANGEMENT UPON THE STEM — NIGHT-BLOOMING CEREUS — EVENING PRIMROSE — EFFECT OF LIGHT UPON THE BLOOM OF FLOWERS — PERFECTING OF THE SEED — THE FRUIT.

Come, brother Freddy, let's go gather some flowers,
   Here are the violets all sweetly in bloom;
And the roses just washed by plentiful showers,
   Will regale with their soft and lovely perfume.

Here are tulips with petals of every hue,
   And a white lily with its bosom so fair;
While daisies and jonquils and hyacinths too,
   Are casting their fragrance around on the air.

The honeysuckles cluster on every spray,
   That twines o'er the lattice or droops from the wall;
Where the Humming-bird sips the nectar away,
   And honey-bees gather their stores for the fall.

Here's sweet flow'ring almonds, a token of spring,
   And yellow corcoras as brilliant as gold;
With the gay Columbine, as pretty a thing
   I'm sure, as we ever need wish to behold.

And primrose and cowslip with poppies intervene,
   Kingcups and primulas all smiling and gay;
Geraniums and foxglove in plenty are seen,
   All standing in bright and imposing array.

Come, while the lark its sweet anthem is singing,
 And the breath of the morn is freshened by showers:
The voice of the thrush through the woodland is ringing,
 Come, little brother, let us gather some flowers.

AMONG the diversified products of Creative Wisdom, there are perhaps no more attractive objects than flowers, and none to which the mind turns with greater pleasure. See how lovely and beautiful they are in their multiplied forms and colors, and how interesting and wonderful in their distribution and uses. Some are decked in colors so brilliant as to bid defiance to all imitation, or marked with tints so delicate as to set at naught the skill of the artist; while others, as emblems of perfect purity, are arrayed in vestures of snowy whiteness.

Nature has scattered these beautiful objects with an unsparing hand over every portion of the globe; they smile in clusters among the decayed leaves of the wood, and the pasture-fields are dotted all over with their ever-varying hues. They rear their gay heads to the sun in gaudy profusion in the ever-glowing regions of the south, and peep out in modest loveliness from beneath the Arctic snows.

There is something happy in the thought that the pleasure to be derived from flowers is open to the youngest, and the poorest of mankind; they are gifts which Nature hands alike to all. It has been said that birds are the poor man's music, so wild flowers may be said to be the poor man's poetry; for him, as for all, they open their gay petals, and exhale the sweetest odors; they smile upon his toils, and add new charms to repose.

To children, flowers are an unfailing source of delight; and the first blossom that flings its fragrance upon the spring air is welcomed by them as a harbinger of future joys. With what care may they often be seen nursing their little daisy-plants, when their whole happiness seems wrapt up in their successful growth! And the violets which they have dug from the woods, and transplanted into their own gardens, are watched with the greatest anxiety. This love of children for flowers is implanted in their young breasts by Him who created every blossom pure and beautiful, and a fit object of admiration and love.

There is much that is interesting and worthy of our attention in flowering plants, besides their beautiful colors, and attractive and showy appearance; many of them possess peculiar habits which render

Complete flower.

Stamens and Pistil.

Ovary and Pistil.

Calyx and Corolla.

Ovary and Calyx.

them objects of wonder. Even the simple parts of a flower, when separated, bear evidence of a superior skill, which has so nicely adapted them to each other. Let us see what they are. First comes the *Calyx*, or the cup which supports the flower; this is sometimes entire, but more frequently parted into divisions, or *segments*, as they are called; it is generally of a pale green color, but, in some instances, as in the Fuchsia, it is highly colored; the Calyx also acts as a covering for the seed-vessels. The delicate and richly colored leaves or *petals*, which stand just within the calyx form the *corolla*. Some flowers have neither calyx nor corolla, and cluster around a pendent spike, as in the Willow and Hazel; these are termed *Catkins*. At the base of the corolla there generally appears the *Nectary*, so called from its secreting a sweet fluid called nectar. This is the store from which the bee derives its honey, and from this delicious fountain the lovely little Humming-Bird, poised upon its rapid wings, extracts through its slender bill the sweet food which it conveys to its young.

Catkin, Hazel.

The most important organs in the flower are those which produce the seed. These consist of two principal parts, called *Stamens* and *Pistils*. They mostly exist in the same flower; but in some cases they not only occupy separate flowers, but are produced upon separate plants. At the base of the pistil is the seed-vessel or *Ovary*, which is composed of one or more valves, differing in form in different plants; a little thread-like stalk called a *Style*, rises from the top of this seed-vessel, and supports a small spongy substance called the *Stigma*. Around this pistil, or pistils, (as there are sometimes many,) are placed the stamens, each consisting of a slender thread, or *filament*, supporting a little bag, called the *anther*, which contains the *pollen*, a kind of powder or dust; when this powder ripens, the anthers burst, and the pollen falls upon the stigma, which is mostly below, and thus the seed in the ovary becomes fertilized. These grains of pollen, which are very minute, when seen under a microscope are of various shapes; some are round or oval, some square, others are toothed like a watch-wheel, or resemble a prickly ball, while others have long appendages or tails.

There is much difference observable in the shape and size of flowers, as well as their colorings; some are large and showy, while others are so diminutive as to require the aid of a microscope to distinguish them. Some are shaped like a bell, as may be seen in the Campanula; others like a trumpet, as the Convolvulus and Honeysuckle; the common Snapdragon and the Scarlet Sage have flowers of a very peculiar form, called *ringent*, or grinning, from their resemblance to an open mouth; but the most common form of flowers is the shape of a star or a cross. They generally consist of from four to eight or ten petals, spreading out like rays, arranging themselves variously; sometimes these petals are broad at the base, and bend upwards, and form a shallow cup; sometimes they bend backwards, and almost clasp the stem; the flower is here said to have its corolla *reflected*.

Flowers also differ in their arrangement. Some grow very close and compact around one common stalk, which is frequently quite long, as in the Foxglove; this is called a *spike*. Sometimes they droop in long and graceful bunches, like Currants; these are styled *racemes*. In the beautiful Lilac they appear in a thick, close head, or *thyrse*. In some cases they hang loosely upon long slender branching stems, or *peduncles;* these are *panicles*, of which the Oats is an illustration. When they have separate stalks which rise from a common centre, and spread out in the form of an umbrella, as in the Carrot, they are described as *umbels;* when these stalks which rise from one centre become much branched, and the flowers more scattered, as may be seen in the common Elder, we call it a *cyme;* if the clusters grow from different parts of the main stalk, and the stems are of different lengths, it is a *corymb;* while if the flowers are on very short stems, and form a close, thick-set cluster, it bears the name of a *fascicle;* of this the Sweet William is a very familiar example.

Spike, Foxglove.

There are also many other modes of flowering peculiar to different plants, but these are the most important, as many of those which come under general observation will be found to have one or another of these methods of displaying their blossoms. There is, however, a very interesting exception to this in the common Dogwood. The flowers, which are quite small, are clustered in close heads, and each head is surrounded by four large white leaves, which are called an *involucre*. These leaves being very prominent and showy, are often mistaken for the flower, while they only act as appendages; but they undoubtedly have some use assigned them; perhaps it may be to protect the delicate little blossoms from the cold night-winds which are apt to prevail in the early Spring, while they are in bloom.

Raceme, Laburnum.

Most flowers require the action of light to cause them to expand, and many never open except under the influence of the most brilliant sunshine. But there are a few instances in which the contrary is observed. Far down in the evergreen forests of South America, when the sun has set behind the tall groves of Palm and Mimosa, and the glimmering twilight is fast following in its train, the magnificent flowers of the Night-blooming Cereus may be seen just opening their fair petals to catch the first rays of the full-orbed moon. Travellers in the tropics describe it as a sight

Fascicle, Sweet William.

worth witnessing, to see in the same forest perhaps hundreds of these lovely blossoms hanging in profusion from the branches of the trees, and loading the atmosphere with the most delicious fragrance. The plants upon which they grow are parasites, and fasten their roots into the

Umbel, Carrot.

trunks and branches of the trees. The flowers are white, and very large, often measuring as much as nine or ten inches in diameter. They commence to blow early in the evening, and remain open during most of the night, when they close, to bloom no more. But the Evening Primrose is a much more familiar instance in which the approach of darkness is hailed by the opening flower. This beautiful and interesting plant grows abundantly in our fields, and on the borders of our woods; and is frequently cultivated in our gardens. It unfolds its pale yellow blossoms in the latter part of the day, and the process of opening is of so remarkable a nature as to claim particular notice. The divisions of the calyx are furnished with little hooks at their extremities, by which the flower is held together before expansion. These divisions open gradually at the bottom, so as to show the yellow corolla within, when suddenly the

Corymb, Candy Tuft.

flower bursts from its confinement, and opens about

half way, being still partially restrained by the calyx; it then continues to expand gradually for some time, when it finally opens with a slight noise. This occupies about fifteen minutes, and may be witnessed upon almost any summer's evening.

Cyme, Elder.

There are also other plants of this description, which are found growing in many parts of the world. The Marvel of Peru has been termed by the French, "Belle de nuit," on account of this peculiarity; and the night-winds of India are laden with the odors of the large blue, lilac, or white blossoms of plants of so magnificent an appearance as to entitle them to the appellation of the "Glory of the night."

Panicle of Grass.

Some plants, the flowers of which bloom many days in succession, close their petals during the night, while in others the leaves double themselves over the blossoms to shelter them from the cold dews. Linnæus, the celebrated Swedish naturalist, termed this "the sleep of plants;" and there is little doubt that nearly all are more or less affected by it, except those whose habits resemble the Primrose. Compound leaves, or such as are composed of many small leaflets arranged on both sides of a common midrib, often fold themselves together, and remain in a drooping posture, until the stimulating influence of the sun's rays causes them again to expand.

While, as has been observed, most flowers require the action of light to make them bloom, the absence of light is not the only cause of their folding up. For although crocuses are so tenacious of their privilege of opening upon the first appearance of the sun, that it is quite easy to cheat them by bringing them near a lamp in the evening, yet many beautiful wild as well as cultivated flowers, regardless of the light, are closed by noonday.

Thyrse, Lilac.

Florists act upon the suggestions of Nature in the management of their choice greenhouse plants; and while they expose them to the full glare of the sun in order to produce the bloom, they also observe that its continued influence tends to hasten decay, by ripening too soon the pollen contained in the anthers, and consequently hastening the fertilizing of the seed; and as the flowers only last in perfection while this process is being accomplished, the period of blooming may be greatly prolonged by shading them from the direct rays of the sun. If, then, the half-opened flower be kept in a sort of twilight by means of canvas or paper shades, the pollen does not

Involucre, Dogwood.

ripen so fast, and the flowers are fair and fresh for many days, and even weeks, instead of yielding to the first brightness of the season. For the moment the great object for which the flower is produced is accomplished, which is the perfection of the seed, it immediately commences to wither, the petals become flaccid, the colors lose their brightness and beauty, and they soon either fold themselves within the calyx, or fall unheeded to the ground. Upon the fading of the corolla, the seed commences to grow, and the ovary which contains it gradually increases until the seed becomes ripe, when it bursts from its confinement, and falls to take root in the earth, and become itself a plant like that which bore it.

There are many curious and interesting forms noticed in the fruits of different plants; some of them have such valuable uses assigned them by man, that without them life would be robbed of many of its luxuries and comforts. The Apples which load our orchard trees, the Peaches and Pears and Plums in almost endless variety, the Grapes and other berries which hang in clusters from our vines, the nuts which lie scattered beneath our forest trees, and above all the grain upon which we depend mainly for our sustenance, are all familiar forms of fruit. How wonderfully does Nature provide, not only for the reproduction of the plant by this means, but how bountifully does she spread around us these her choicest blessings, which are so singularly adapted to our wants!

The leaves also of plants present many varieties, both in their shape and arrangement. Sometimes they are placed alternately one above another on the stem; sometimes two are placed opposite each other; and often we see them in what is called a *whorl*, or radiating from the same point like the spokes of a wheel. They also occur in tufts or bunches thickly scattered on the stem or branches, and sometimes but a single leaf is seen, and that springs immediately from the root, and is termed a *radical*, while those which grow from the stem are called *cauline*. Some plants have both cauline and radical leaves, and some have neither.

The following cuts will illustrate the principal shapes observable in leaves.

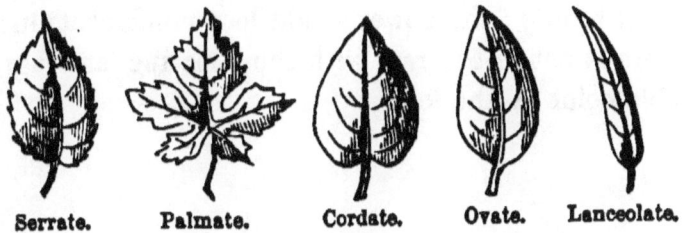

Serrate.   Palmate.   Cordate.   Ovate.   Lanceolate.

These may be separated into two distinct classes, the simple and the compound; the simple being those

FLOWERS.

Sagtitate.   Sinuate.   Mucronate.   Digitate.   Crenate.

which, though much notched, are not divided into separate parts. The Fuchsia has a simple leaf. The compound are such as consist of a number of small leaflets arranged upon a common midrib, as is seen in the Sweet Pea.

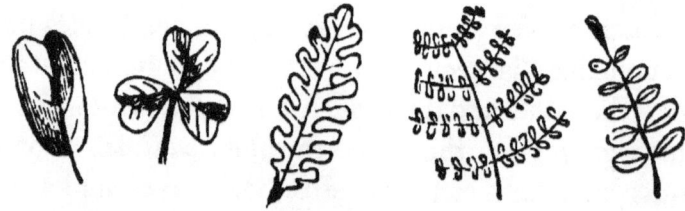

Emarginate.   Ternate.   Pinnatifid.   Bipinnate.   Pinnate.

Leaves may be considered as the most important appendages of plants, and certainly add not a little to their beauty; the flower would lose much of its lustre were it not in contrast with the pleasing and agreeable color of the leaf.

CHAPTER III.

THE AMERICAN ALOE — THE PASSION FLOWER — ORCHIDS — TRUMPET FLOWER — MORNING GLORY — THE ROSE — BLACKBERRY — DAISY — COMPOUND FLOWERS.

OF all the forms in which flowers appear, there is perhaps none more wonderful than that of the Aloe. Its peculiar habits, and its gigantic dimensions, may well entitle it to the name of king of flowers.

It is commonly known by the name of "The Century Plant," from the fact that it was formerly supposed to bloom only once in a hundred years. This is, however, an error which time has corrected, as many specimens have been known to flower in conservatories in much shorter periods; and it is probable that in its native climate it occurs at an early age. In the United States the Aloe is probably the best known, and most frequently kept as an ornament to our hot-houses. It is a native of tropical America, where it is a plant of great utility to the Indians.

The singular fact that it blooms but once, and that its existence terminates with the decay of its flowers, has rendered it particularly interesting; and as the opportunity of witnessing so beautiful a spectacle is of rare occurrence, it is regarded as a great curiosity

American Aloe.

A noble specimen, probably 85 or 90 years of age, recently bloomed in Philadelphia About the middle of the summer of 1858, the stem made its appearance, and in six weeks' time it had reached the height of about twenty feet, being seven inches in diameter at the base, and crowned with seventeen fascicles of greenish yellow flowers, numbering in all about 3000, and spreading over a space of nearly eighteen feet in circumference.

In Mexico, the West Indies, and South America, where several varieties of this plant are found, it is often cultivated by the natives, and its different parts appropriated to useful purposes. It may frequently be seen planted in long rows, which serve as hedges, and form an impervious barrier both to man and beast.

In no other country, perhaps, is the Aloe so generally serviceable as in America.

The rope bridges of Mexico, so often named as dangerous to the traveller unaccustomed to cross them, are formed entirely of cords made of the fibrous parts of its root. These bridges, swung over some foaming torrent, have pieces of the bamboo stem placed at small intervals across the ropes, disclosing through their interstices the dashing of the waters; and their rude structure, oscillating either with the wind or the unsteady footsteps of the passengers, might appal the heart of the strongest and bravest strangers, while the Indian passes lightly and fearlessly over.

The leaves of the Aloe, when baked, form an excellent article of food, and from the juice, sugar and medicines are prepared. The strong flower stems serve as beams for the roofs of the Indian dwellings, and the leaves supply the place of tiles.

In former times the Aloe was extensively cultivated for the manufacture of paper, and great quantities were evidently used in the time of the Montezumas in painting hieroglyphics. The paper produced from this plant resembles that made by the Egyptians from the papyrus. The ancient Mexican manuscripts, which have received so much attention from the learned, and upon whose records is based the history of that injured and interesting people, were painted chiefly upon paper made from the fibre of the aloe. Many of these "picture writings," as they have been called, are still preserved at Mexico.

From the juice pressed from the flowers of this plant, the natives prepare a very pleasant and refreshing beverage, called "pulque," of which they are very fond, and it is said to be quite nutritious and wholesome, although, if taken in large quantities, it produces the same effect as brandy.

The drug called aloes is the thickened juice of a species of aloe, which grows abundantly near the Cape of Good Hope. It is procured by cutting the leaves in pieces, and pressing and boiling them; after which the juice is allowed to cool, when it becomes hard.

How few, who look upon the thick spiny leaves of the Aloes, as they stand in our green-houses, ever reflect upon the great usefulness of this plant to the natives of America!

Hanging in gay festoons about the venerable trees of the American forests, the various kinds of Passion-Flowers form objects of splendor which arrest the attention of the traveller. In this, their native soil, they grow to much greater perfection than when kept in our green-houses; and their large starry blossoms hang down in profusion among the branches, or clasp by their strong tendrils the immense trunks of the trees.

There are upwards of forty species found growing in various parts of the world, varying greatly in their color and appearance. Some are very strong and robust plants, sending out long stout stems which lay hold of anything within their reach; and in the summer season, when their growth is rank and rapid, they soon envelope the trunks of the trees in a rich and luxuriant mantle. These have mostly large blue, white, or crimson flowers, which they bear in great abundance. The greatest number of varieties may probably be found in South America and the West Indies. One or two species grow in the United States, and many in Africa and the adjacent islands.

These flowers are of short continuance, generally lasting but one day, opening a little before noon, and closing in the evening.

The name of Passion-Flower was given to it by the Spaniards, whose attention was attracted by the beautiful and showy appearance made by the vines in the forests of Mexico and South America; and fancy pictured to them in the various parts of the flower a resemblance to the crown of thorns, and the other signs of our Saviour's passion. Alluding to this, the poet Barton says:—

> "We soar to heaven; and to outlive
>   Our life's contracted span,
> Unto the glorious stars we give
>   The names of mortal man.
>
> Then may not one poor floweret's bloom
>   The holier memory share
> Of Him who, to avert our doom,
>   Vouchsafed our sins to bear?
>
> God dwelleth not in temples reared
>   By work of human hands;
> Yet shrines august, by man revered,
>   Are found in Christian lands.

And may not e'en a simple flower
  Proclaim his glorious praise,
Whose fiat only, had the power,
  Its form from earth to raise?

Then freely let the blossom ope
  Its beauties — to recall
A scene which bids the humble hope
  In Him who died for all!"

Butterfly Orchis.

In the same countries where the Night-Blooming Cereus and the Passion-Flower wreathe their bright blossoms among the forest-trees, may be seen many other curious and interesting plants; among them is a tribe known as Orchises or Orchids; these, like the Cereus, are often found growing upon the trunks and branches, and sometimes in such quantities, as almost to deprive the tree of the nourishing sap intended for its support. If we should attempt to describe the multiplied forms and colorings of these air-plants, it would cost the labor of a lifetime. They mostly consist of rough unsightly bulbs, which, for about one-half the year, lie apparently lifeless, adhering by their tough fibrous roots to the bark; but no sooner do the clouds of the rainy season empty their life-giving virtues upon the earth, than they send out their leaves in thick tufts, which, being often long and grass-like, have a graceful appearance. After these have arrived at their full perfection, the flower-stems shoot forth, and are sometimes several feet in length, loaded with a profusion of gay flowers, frequently very large, and of almost every imaginable shape and color. Some of them bear a close resemblance to living objects. Thus, in the Butterfly Orchis, the likeness is so striking, that one unacquainted with the plant would suppose that a large yellow butterfly had chanced to light upon it.

Orchises are divided into two kinds, terrestrials or such as grow upon the ground, and Epiphytes, or those which hang from the trees. Many very beautiful terrestrial Orchids are to be found in our own woods and meadows. But the most curious and remarkable species are exotics, and require peculiar treatment to make them flower. The roots are tied to a rough stick of wood, with the bark upon it, and are suspended from the roof of the green-house; others are planted in pots filled with stones and rotten wood. The air in the house being kept constantly moist, the plants thrive and produce their blossoms in great perfection.

Another very attractive and showy species of plants is the Bignonia, or Trumpet-Flower; of this, many varieties abound in the same localities as the Orchids, and may often be seen weaving a tangled web with their long twining stems as they clamber over the trees. The great abundance of these and other vines in the tropical forests, so fills up the recesses between the undergrowth, as to render them almost impenetrable, and travellers often speak of being compelled to cut their way through with axes. These tangled brakes are the lodging-places of thousands of beautiful birds, which build their nests and rear their young without fear of intrusion. Here the jewelled breast of the Humming-Bird is seen glittering in the light, as it flits from flower to flower and gently dips its tiny bill into their sweet ambrosial cups; and here may be heard the wild screams of flocks of gay plumaged Parrots, intermingled with the no less clamorous chattering of troops of monkeys.

On the borders of these forests may also be found in great profusion, many elegant varieties of Convolvulus or Morning-Glories, plants with which most persons are familiar. Of all the flowers with which Nature with a lavish hand adorns our gardens, there is perhaps none more showy or more fleeting. Their delicately painted petals, their luxuriant growth, the graceful drooping of their long twining branches, and, above all, the gorgeous array of large showy blossoms, which welcome with their smiles the earliest streaks of dawn, all conspire to render them, as they truly are, the glory of the morning. But how instructive is the lesson conveyed by the language of Scripture, which is so applicable to this beautiful plant:—" For the sun is no sooner risen with a burning heat, than the flower thereof falleth, and the grace of the fashion of it perisheth."

The common sweet potato is a species of Convolvulus, and was originally brought from South America. Its blossoms are bright purple, but are so hidden beneath the leaves, as to attract but little attention.

Most varieties of Convolvulus, or Ipomæa, are annuals, and are grown from seed, but there are some whose roots are perennial, and in their native climates they are constantly clothed with verdure. One of these, which, in this latitude, requires the protection of the green-house during the winter, extends its branches to a great length, sometimes forty feet in one season, and over four hundred flowers have been counted at one time upon a single vine, each flower being four inches in diameter.

But foremost among the flowers of the garden stands the rose, a general favorite on account of its soft and delicate colorings, and its delightful fragrance.

The rose was undoubtedly well known, and its qualities appreciated, at a very early period. In the Scriptures, it is alluded to, where the idea of great beauty and excellence is intended to be conveyed. Solomon, in Canticles, speaks of the "Rose of Sharon;" and the prophet Isaiah, in ch. xxxv. 1, thus makes use of it in a beautiful comparison: "The wilderness and the solitary place shall be glad for them, and the desert shall rejoice and blossom as the rose."

We are greatly indebted to cultivation for the perfection in which the many beautiful varieties of this lovely flower now exist. Most of our finest roses were originally brought from the East, where they are raised in large quantities as an article of commerce. Ghazipore may be called the rose-bed of India. In the spring of the year, an extent of miles around the town presents to the eye a continued garden of roses. The sight is perfectly dazzling, the whole surface of the ground, as far as the eye can reach, being clothed with the same beautiful carpet of mingled green and red; while the air is loaded with the sweetest odors, which are wafted far across the river Ganges. The flower is cultivated thus extensively for the manufacture of rose-water.

There is much interest connected with the cultivation of this flower; the almost indefinite number of sorts, with hues varying from the most delicate pink to the deepest crimson, and from the purest white to a brilliant yellow, renders it a peculiarly fit object to adorn our conservatories or our flower-gardens. And while its blossoms are proverbially frail, and continue at most but a few days, still its rapid and constant succession of fresh opening buds fill up the places of those which have fallen beneath the rays of the sun.

> "Fairest flower, the pride of spring,
> Blooming, beauteous, fading thing
> 'Tis as yesterday, when first
> Forth thy blushing beauties burst,
> And I marked thy bosom swell,
> And I caught thy balmy smell,
> Fondly hoping soon to see
> All thy full-blown symmetry:

But I came — and lo! around
Sadly strewn upon the ground,
Lovely, livid leaves I see —
Oh! can these be all of thee?
I could weep, for so I've known
Many a vivid vision flown;
Many a hope that found its tomb
Just when bursting into bloom;
Many a friend — Ah! why proceed?
See afresh my bosom bleed —
Rather turn my thoughts on high,
Hopes there are which cannot die;
Yes, my Saviour, thou canst give
Joys that will not thus deceive;
Eden's roses never fade,
Eden's prospects have no shade."

There are some beautiful kinds of roses found wild in almost every part of the country; when unaltered by culture they are generally single, and have but five petals, with a great profusion of stamens, which fill up the space between the pistil and corolla. In the cultivated varieties, the number of the petals is greatly increased, while the stamens are not so numerous. The common blackberry belongs to the same class of plants as the rose, and if the peculiar formation of the flower and leaf be compared with that of the wild-brier or rose, the resemblance will at once be seen.

The daisy is so well known to all, that any description of it would seem useless; but, as there are several different plants known by that name, it may be well to speak of them. The bright little flower that is so welcome in the spring, is the common daisy of Europe; and it is this lovely plant that has formed the theme of many a poet's song. Wordsworth's three beautiful poems are too familiar to be quoted. Spenser sang of the "little daisie that at evening closes;" while Chaucer and Ben Jonson each had a good word for the bright "day's eye."

What is commonly called the daisy in this country is a species of Chrysanthemum: it is also of foreign origin, and is one of those plants whose beauty hardly repays for the trouble it gives the farmer, as it increases so rapidly, both by its roots and seed, that where it once obtains a footing, it soon spreads over whole fields, thus preventing the growth of that which is more valuable. It is considered by the Danes to be so injurious to the pasture, that one of the laws of Denmark compels the farmers on whose land it appears, to use every effort to eradicate it.

There is also another variety of plants to which the name of Michaelmas daisy has been applied in England; they consist of various kinds of asters, some of which have been introduced there from America and China. These are among the last flowers of summer, even blooming until late in the autumn. Some of them are possessed of great beauty; the well-known China Aster, or Queen Marguerite, is among their number.

The flowers of the daisies are what are termed compound, or similar to those of the dog-wood. The beautiful white, blue, or rose-tinted petals, which are so conspicuous, are the rays of the involucre, and it is in the centre of these where the compact mass of minute flowers lies; so that what is commonly regarded as one, consists sometimes of hundreds of blossoms. Each of these being a complete flower, with its corolla, pistil, stamens and seed-vessel, as may readily be seen under the microscope.

Many beautiful compound flowers ornament our gardens during the summer and autumn. The stately sun-flower, which grows to an immense size in the woods and plains of Mexico, and excited the astonishment of the Spanish conquerors; the bright marigolds, some of which bloom in almost every month of the year; the dahlias and chrysanthemums, with those dear friends the daisies, which welcome the spring, and around whose quaint little name so many pleasant associations cluster.

> 'Not worlds on worlds, in phalanx deep,
>   Need we to prove a God is here;
> The daisy, fresh from Nature's sleep,
>   Tells of his hand in lines as clear.
>
> For who but He who arched the skies
>   And pours the day-spring's living flood,
> Wondrous alike in all he tries,
>   Could raise the daisy's purple bud;
>
> Mould its green cup, its wiry stem,
>   Its fringed border nicely spin,
> And cut the gold-embossed gem
>   That, set in silver, gleams within;
>
> And fling it, unrestrained and free,
>   O'er hill and dale, and desert sod,
> That man, where'er he walks, may see
>   In every step the hand of God?"

CHAPTER IV.

LILIES—VICTORIA REGIA: ITS DISCOVERY—LOTUS OF EGYPT—LILIES MENTIONED IN SCRIPTURE—TULIPS—CALLA LILY—JACOBEAN LILY.

"Observe the rising lily's snowy grace,
Observe the various vegetable race;
They neither toil, nor spin, but careless grow,
Yet see, how warm they blush! how bright they glow!
What regal vestments can with them compare!
What king so shining! or what queen so fair!
If ceaseless thus the fowls of heaven He feeds,
If o'er the earth such lucid robes He spreads;
Will He not care for you, ye faithless, say?
Is He unwise? or, are ye less than they?"

OW many are the charms which hang around this lovely and beautiful flower, of which it has been said by the great Redeemer, that "Solomon, in all his glory, was not arrayed like one of these!"

As an emblem of purity, some species are peculiarly appropriate, their snow-white petals being unsullied by a stain; while others are decked in all the rich and glowing colors of the rainbow.

LILIES. 51

The principal mention of the Lily in Scripture, is in Canticles, where Solomon frequently uses it as conveying the impression of great beauty and loveliness.

There are many flowers to which the name of Lily is applied, between some of which there seems to be but little resemblance. Among these, perhaps there is none more beautiful, and certainly none more wonderful, than the Victoria Regia, or the Great Water Lily. No description of this remarkable plant can convey to the reader any adequate idea of its singular beauty. It stands alone among its tribe as Queen of the Waters, nothing having yet been discovered which can compare with it. (See Frontispiece.)

This plant, although known to naturalists as early as 1822, was not introduced into England until about the year 1846, when seeds were taken thither by Thomas Bridges, a collector of curious plants in South America. We extract the following narrative from the published account of his discovery.

"During my stay at the Indian town of Santa Ana, in the province of Moxos, republic of Bolivia, during the summer of 1845, I made daily shooting excursions in the vicinity. In one of these I had the good fortune, whilst riding along the woody banks of the river Yacuma, one of the tributary rivers of the Mamoré, to come suddenly on a beautiful pond, or rather small lake, embosomed in the forest, where, to my delight and astonishment, I discovered for the first time, the Queen of Aquatics, the Victoria Regia! There were at least fifty flowers in view, and Belzoni could not have felt more rapture at his Egyptian discoveries, than I did in beholding the beautiful and novel sight before me, such as it has fallen to the lot of few Englishmen to witness. Fain would I have plunged into the lake to have procured specimens of the magnificent flowers and leaves; but knowing that these waters abounded in alligators, I was deterred from doing so by the advice of my guide, and my own experience of similar places.

"I now turned over in my thoughts how and in what way flowers and leaves might be obtained; and I clearly saw that a canoe was necessary, and therefore returned promptly to the town, and communicated my discovery and wants to the Correjidor or Governor, who with much kindness immediately ordered the Cacique to send Indians with a yoke of oxen for the purpose of drawing a canoe from the river Yacuma to the lake. Being apprised that the canoe was in readiness, I returned in the afternoon, with several Indians to assist in carrying home the expected prize of leaves and flowers. The canoe being very small, only three persons could embark; myself in the middle, and an Indian in the bows and stern. In this tottering little bark we rowed amongst magnificent leaves and flowers, crushing unavoidably some, and selecting only such as pleased me. The leaves being so enormous, I could find room in the canoe but for two, one before me and one behind; owing to their being very fragile, even in the green state, care was necessary to transport them; and thus we had to make several trips in the canoe before I obtained the number required. Having loaded myself with leaves, flowers, and ripe seed-vessels, I next mused how they were to be conveyed in safety; and determined at length upon suspending them on long poles with small cords, tied to the stalks of the leaves and flowers. Two Indians, each taking on his shoulder an end of the pole, carried them into the town; the poor creatures wondering all the while what could induce me to be at so much trouble to get at flowers, and for what purpose I destined them now they were in my possession."

The leaves of this plant are round, and vary considerably in size, the largest being about six feet in diameter. They float on the surface of the water, are of a light green color above, and bright purple below. The margins of the leaves are turned upwards, giving them the appearance of floating dishes. The plant grows in four to six feet of water, producing leaves and flowers which rapidly decay and give place to others. From each plant there are seldom more than four or five leaves on the surface; but even these, in parts of the lakes where they grow most abundantly, almost cover the surface of the water, one leaf touching the other. The blossoms rise six and eight inches above the surface, expanding first in the evening, when they are pure white, changing finally to a beautiful pink or rose color; and several may be seen at the same time, partaking of every tinge between the two. The largest flowers measure about ten or twelve inches in diameter.

Another very interesting and beautiful plant, similar to the above, although much less in size, is the Lotus, or the famed Egyptian Water Lily, which was formerly adored as a deity by the idolatrous inhabitants of that country. It also grows in the river Ganges as well as the Nile, and is held in the same veneration by the natives of Hindostan and Nepal. One of the latter, upon entering the study of Sir William Jones, prostrated himself before some specimens which happened to lie there for examination. The Egyptians prepare a kind of bread from its seeds, and sometimes feed upon its tuberous roots.

Pond Lily.

The Lotus resembles very closely our own White Pond Lily, except that the flowers and leaves, instead of resting upon the surface of the water, rise some distance above it.

Some authors believe the Lotus to be the Lily of the Old Testament, as very frequently, in Canticles, Solomon speaks of his beloved "feeding among lilies;" and the root, stalks, and seeds being common articles of Egyptian diet, would naturally lead to such a conclusion, especially as it is supposed that the

Song of Solomon was written on the occasion of his marriage with an Egyptian princess.

White Lily.

The great beauty of the common White Lily of our gardens, would naturally suggest that it was the one so often spoken of in Scripture; but as it is not certain that this was a native flower in Palestine, it seems more probable that the plant often referred to is, as Dr. Kitto believes, the Yellow Amaryllis, which covers large tracts of country in the Holy Land, and blooms until so late in the year, as to be almost in its prime when most other plants have yielded to the influence of the cold.

The many-colored Tulip, whose gorgeous tints would outshine even the robes of eastern royalty, has also been supposed by some to be the subject of our Saviour's allusion when he so beautifully and so tenderly encouraged the drooping faith of his disciples. The beauty of the Tulips in the plains of Sharon, as well as at Joppa, has frequently attracted the notice of British travellers; and even the gayety and brilliancy of a bed of Tulips in our own gardens, is an unfailing source of admiration.

Another kind of Lily which adorns our greenhouses and gardens early in the Spring, is the Calla. This plant is a native of Africa, and frequents low, wet grounds, where its tall leaves and flower-stalks stand two and three feet above the water, the latter bearing a long spike of flowers, surrounded by one beautiful broad petal or sheath, of the purest white, this sheath is termed a *spadix*, and forms a good example of a very singular mode of flowering.

The Jacobean Lily is a species of Amaryllis, of a deep crimson or almost mahogany color, and flowers also in the Spring. There is a very curious process by which the seed becomes fertilized in this plant. In the morning a drop of very clear liquid issues from the stigma; this liquid receives the pollen which falls from the anthers, and soon becomes thick and turbid, and about noon is so heavy as to be almost ready to drop, when it is again absorbed, carrying with it the fertilizing principle of the stamens.

These plants, though all known by the familiar name of Lilies, belong to entirely different families; the true Lilies embracing only those which are not aquatics.

THE CHILD AND THE LILY.

I saw, one morn, a little maid
  With locks of golden hair,
Pluck from its stem beneath the shade
  A lily bright and fair.
And with a heart all full of glee,
  "Oh! dear mother!" she cried,
"Look what a sweet charm this will be
  To set here by my side;
For now I'll smell its soft perfume,
  Its graceful form will view;
And gaze upon its placid bloom,
  All decked with shining dew.

Oh! can it be that here below,
  All o'er the verdant plain,
This fair and beauteous flower should grow
  And bud, and bloom, in vain?
It looks so sweet, and pure, and good,
  Within its robes of white,
It makes me wish that if I could,
  I too might look so bright."

"Oh! then, my child, if thou wouldst be,"
  The mother soft replied,
"Like that fair flower from spot so free,
  Or taint of earthly pride,
Lift up thy heart to God above,
  Who reigns supreme on high;
And ask, that in His matchless love
  He'd deign to hear thy cry;
And from thy soul to wash away
  Each foul and guilty stain,
And on thy spirit shed a ray
  Of life and peace again.
Ask that thus washed thy robes may be,
  Pure as the lilies fair;
That thou, from sin forever free,
  Christ's spotless robe may wear.
And let thy youthful heart be riven
  From this vile world away;
And all thy hopes be fixed on heaven,
  The realms of endless day;
For there, within His fold of rest,
  Amid unfading light,
The ransomed soul, forever blest,
  Shall walk with Him in white.'

CHAPTER V.

SIMILARITY BETWEEN THE FUNCTIONS OF PLANTS AND ANIMALS—THE AQUARIUM—PRINCIPLES UPON WHICH IT IS SUSTAINED—EARLY EXPERIMENTS WITH THE AQUARIUM—PLANTS MOST SUITABLE FOR THE PURPOSE—SEA WEEDS, MOULD, LICHENS, MOSSES, FERNS.

T is a very curious fact that in many of the functions of plants, we observe a close resemblance to those witnessed in animal life; thus, the circulation of the sap, which will be more fully described hereafter, is in effect precisely similar to the circulation of the blood in the human body, vessels being provided in each, which are peculiarly adapted to carrying the fluids which support their existence to the parts where they are needed. Respiration is also a point in which great similarity exists. Leaves are the breathing organs of plants; through them the sap is brought into contact with the air, where it absorbs that which is necessary for its purification. In this operation we shall notice a wonderful provision by which nature seeks to preserve a proper balance between the requirements of the animal and vegetable world. In the purification of the blood, the air taken

The Aquarium.

into the lungs is deprived of a large amount of oxygen gas, while at the same time it becomes charged with carbonic acid gas, which is incapable of supporting animal life. This noxious principle is absorbed by the leaves of plants, where it appears to undergo decomposition; the carbon being retained for the use of the plant, and the oxygen liberated to assist in restoring the atmosphere to its original purity. This action takes place only under the influence of light, as during the night the contrary occurs, the leaves giving out carbon and absorbing oxygen, although in very small quantities compared with what is emitted during the daytime.

If a bunch of leaves be introduced into a jar of air which has been deprived of its vitality by means of animal respiration, and the jar exposed to the rays of the sun, the air will, in a few hours, again become pure and wholesome.

The same principle holds good in aquatic plants, many of them having the power of keeping the water in which they grow from becoming impure or foul.

In ponds where there is no regular supply of fresh water from running streams, it has been noticed that, during the winter, when the plants are dead, the fish frequently come to the surface to breathe, while in the summer, when the plants are growing, the vitality of the water is preserved.

It is upon this wonderful law of Nature that the aquarium, that endless source of amusement and instruction, is based; and although it is as much intended to illustrate the functions of animal as of vegetable life, perhaps the following account, taken from a beautiful work, entitled "Ocean and River Gardens," descriptive of the principles upon which it is conducted, may be interesting to the reader.

"The successful treatment of aquatic plants and animals, in the confined space of a glass aquarium, depends entirely upon the discovery that there exists in Nature a self-adjusting balance between the supply of oxygen created in water, and the quantity consumed by aquatic animals. And it became equally necessary to know the means by which that supply was continually generated. Without the knowledge of these facts, and the principles by which they are regulated, it would have been impossible to establish such a marine aquarium as we may now any day examine in the Regent's Park (London); where, in a few glass tanks, of very moderate size, we may see examples of some of the most curious forms of animal and vegetable life peculiar to the depths of the ocean; forms so singular, that their first exhibition created a sense of wonder little less intense than that which must have been caused, long years ago, by the first public display of the mountain form of the Elephant to the people of cold northern countries.

"Those principles, the knowledge of which was requisite to enable us thus to view the wonders of the Ocean in their living state in the aquarium, were not mastered at once, or by one man, or in one generation. The nature of certain relations between animal and vegetable life, upon which they are founded, was first advanced by Priestley, towards the close of the last century, who proved that plants give forth the oxygen necessary to animal life.

"But it was not till the year 1833, that Professor Daubeny communicated to the British Association at Cambridge, a paper concerning some new researches prosecuted in the same direction; while in the summer of 1850, R. Warrington communicated to the Chemical Society a series of observations on the adjustment of certain relations between the animal and vegetable kingdoms, very important to our present purpose. Two small gold-fish were placed in a glass receiver, a small plant of Valisneria Spiralis being planted at the same time in some earth, beneath a layer of sand in the same vessel. All went on well by this arrangement, without any necessity for changing the water; the oxygen given off by the plant proving itself sufficient for the supply of its animal co-tenants, and the water therefore remaining clean and pure, until some decayed leaves of the valisneria caused turbidity. To remedy this evil, he brought to bear the results of previous observations on water in natural ponds under analogous circumstances; and, guided by these observations and their results, he placed a few common pond-snails in the vessel containing his gold-fish and plant of valisneria.

"The new inmates, immediately upon their introduction, began to feed greedily upon the decaying vegetable matter, and all was quickly restored to a healthy state. They proved, indeed, of still further advantage, for the masses of eggs which they deposited evidently presented a kind of food natural to the fishes, which was eagerly devoured by them, so that the snails became not only the scavengers, but also the feeders of the little colony. And so this first of true aquaria prospered; the animals and plants proving of mutual value and support to each other.

"By the culture of some of our most beautiful fresh-water plants, in glass aquaria, many of the wild beauties of Nature, in some of her most pleasing and interesting aspects, may be wrought into attractive decorations for our ordinary living-rooms, with very little trouble or expense.

"By means of an aquarium, the forms and habits of fish, reptiles, and aquatic insects [also,] may be made to develop themselves under our eyes, undisturbed by the continual necessity of changing the water; thus affording us the curious spectacle of many phases of animal life that have hitherto been concealed in depths inaccessible to the observation of the most curious."

A very interesting circumstance which appears to have occurred during some of the early researches of the same author in aquatic animal life, although a digression from our subject, is too curious to be omitted.

He says, "A strange, scorpion-like creature, after exercising its voracious appetite upon every other living thing in the vessel in which I had placed it, seemed suddenly to lose all taste for the luxuries of the palate, notwithstanding a copious supply of the living delicacies it was most fond of, and with which I had taken care to furnish it at regular intervals. It became restless and apparently diseased, and I concluded that I was about to lose this favorite specimen as I had lost so many others. Its uneasiness, however, took quite a different turn to the one I expected, ending in nothing less than a determination to leave its native element. Had I seen a Carp or a Tench quietly walk out of the fish-pond and climb a tree, I could not have been more astonished than when I saw this creature of the water, which, with its fin-like tail, and other appendages, was evidently intended for a denizen of that element, quietly crawl up a stick which was standing in the vessel, and, emerging from the water, remain quietly attached to the support it had selected, at some inches above the surface of the element it thus so strangely and suddenly quitted. Its determination appeared the more astonishing, as I soon perceived its finny tail, its legs, and at last the whole of its skin, gradually hardened and blackened, and it appeared to have shared the natural fate of a fish out of water. After watching it for some days, without perceiving any further change, other matters occupied my attention, and I entirely forgot the fate of my voracious pet, which had met such an untimely end in consequence of rashly leaving the proper sphere of its existence.

"Some little time afterwards I was about to empty the jar, and throw away the stick to which the dried and hardened form of the victim to getting out of bounds was still attached, when I thought I perceived a division in the blackened skin of the back. As I saw that the opening widened, my curiosity became again excited, and I determined to watch and see if any other change would follow. Taking a book, therefore, I sat down near the object of my attention. I had not read many pages, turning frequently towards the remains upon the stick, when suddenly— I shall never forget the surprise of the moment— when suddenly the opening of the back was much widened, as by some sudden effort, and the greater part of a glittering Dragon-fly became plainly visible; very quickly the whole insect emerged from the blackened shell, spreading its great gossamer wings to the sun, which was shining brightly through the window.

"I had by an accident, for I can hardly call it the result of a course of observation, witnessed one of the most extraordinary and complete of the metamorphoses that occur in the whole range of insect life, and was all anxiety to pursue my discoveries. I was, however, baffled in all future attempts, at that time, to extend my knowledge of the mysterious creatures of the world of waters; and it was not till recent discoverers have shown how the Aquarium may be made the means of facilitating studies of that class, combined with an elegant and delightful mode of amusement, that I resumed the course of observation which has been so long interrupted by difficulties which appeared insurmountable."

Those plants which naturally grow entirely below the surface of the water, are best calculated for the purpose of the Aquarium, as they are less liable to decay; and their leaves being mostly very fine and delicate, they not only present a more beautiful appearance, but the breathing organs are more generally distributed throughout the water. This is particularly noticeable in marine plants, they consisting frequently of bunches of delicately formed filaments, of so fragile a texture as to be very easily broken, but which float at their ease upon the ocean, waving to and fro with the motion of the water as gracefully as the trees wave before the winds.

In the illustration on page 59, the tall and graceful form of the Calla will be recognized rearing its beautiful flowers far above the surface of the water, while below will be seen the forms of some of the most interesting aquatics.

The foliage of the Myriophyllum presents a fine appearance when seen floating in the water, the very minute divisions of which have given it the name of Milfoil, or thousand leaves.

The Water Buttercup is also a very interesting plant, on account of its peculiar growth; the leaves which appear below the surface of the water are so deeply cut, as apparently to consist of nothing but veins or fibres, while those which are developed above are broad and flat, the veins being connected by the ordinary tissue. The Starwort also presents the same formation; the foliage below is long and slender, while it spreads out upon the surface in beautiful whorls, somewhat like a star. Here we see peculiarities adapted to two different elements, existing in the same plant.

The number of plants which may be grown successfully in an Aquarium, is great; but for ordinary purposes, three or four well-selected varieties are sufficient. In all cases a specimen of Valisneria Spiralis should be obtained, if possible, as its grass-like appearance is particularly appropriate, and it is an excellent generator of oxygen.

The flowing and delicate forms, and the richness of coloring of many of the "Sea Weeds," as they are called, render them objects peculiarly worthy of our attention. The careless lounger at the sea-side, as he casts his vacant gaze over the swelling bosom of the deep, dreams not of the store of hidden treasures which lie veiled beneath its waters. Little does he think that wave upon wave, as they roll in ceaseless succession, tossing their snowy crests upon the pebbly shore, come freighted with the beauties of many a far-distant clime. But an eye accustomed to recreate among the varied scenes which adorn this beautiful world, cannot but feel an irresistible longing to lift the folds of that broad curtain which separates him from the wonders of the vast mysterious ocean. Each new-born gale that wafts its saline fragrance o'er the white-capped billows, and every ripple that laves his feet, is laden with themes for suggestive thought; while every tide that flows, bearing upon its swell jewels from the profoundest depths, itself unveils in its ebbings the beauties of a "world beneath the sea."

Sea Weeds.

If you will examine the beach during the recess of the tides, particularly after a storm, you will find it thickly strewn with fragments of the most beautiful plants; some being colored with the most brilliant shades of crimson, some sparkling as with gold, or glittering like silver, and all possessing a very peculiar and curious formation; while in the little pools among the rocks may be seen many of the lower forms of animal life, which are truly wonderful. Almost any of the marine plants are suitable for the Aquarium, and it is here that their peculiar habits may be most carefully studied. They generally thrive well with little care, and mostly present a singular appearance, fastened to the rocks, and growing we scarce know how. A few of these, well chosen and tastefully arranged in a glass tank, together with shell-fish, Sea Anemones, and a few Sticklebacks and Minnows to give life to the whole, will form an object which cannot fail to interest the most unthinking individual.

There are some plants found growing on the rocks near the sea, which, although they resemble the sea-weeds in some respects, belong to a different class, and a slight knowledge of botany will enable any one to distinguish between them.

The Samphire is an example; it is an umbelliferous plant, and never grows below the surface of the water, but fastens itself upon the rocks just beyond the reach of the tide, but where it can receive sufficient moisture from the spray.

An interesting anecdote is related of some shipwrecked mariners who owed the preservation of their lives to the knowledge of the habits of this plant, possessed by one of their number.

It was many years ago that a large ship was driven upon the rocks in the English Channel, upon which she soon became a wreck. The entire crew were lost except four, who clung to a large projecting crag, which appeared to be the only refuge to which they could resort. The darkness of the night rendered every other object invisible, except when the vivid flashes of lightning would cast upon the wild scene around them a momentary glare, revealing the true horrors of their forlorn condition. This was rendered the more hopeless as they perceived that the tide was rising, and the spot on which they stood was decreasing in size as each succeeding wave broke over them. The storm was too violent to admit of their being heard from the shore, and the melancholy thought that they would soon be driven from their only hope of safety by the advancing waters was truly disheartening. Just at this moment, when they were debating whether or not they should commit themselves to the mercy of the waves, in hopes of reaching some more elevated position, one of them, while endeavoring to hold more firmly to the rock, grasped a weed, which, wet as it was, he at once recognized as the Rock Samphire, which he knew never grew beneath the water. The knowledge of this fact, indicating that the tide had nearly reached its highest point, assured them that they might remain with safety. Their anxiety was at once relieved, and the rest of that dreadful night passed in comparative comfort. At daybreak their perilous condition was discovered from the shore, and they were rescued "A little learning," in this case, was certainly no "dangerous thing."

The Sea Weeds, or marine "Algae," as they are termed, belong to the first great natural order of plants, — the Acrogenous; they are so called because, with a few exceptions, they are devoid of the usual appendages of plants — stems, leaves, and flowers. Some of the simplest forms belonging to this order consist merely of a mass of cellular tissue. The mould which collects in damp places, and sometimes upon the top of articles of food that have been kept in damp closets, is a little plant of this order. The green tinge assumed by stagnant water, is owing to the presence of a species of fresh water "Algae," which grows spontaneously in such places. The beautiful lichens that cover the bark of some trees, and the rails and boards of old fences, the many kinds of moss with which our woods abound, and the unsightly mushroom and toadstool, all belong to this order of plants. In all these there exists nothing which can be strictly defined as either stem, leaf, or flower; but in the "Ferns," which also belong to the same order, we see the connecting link between the higher and the lower forms of vegetable life. The

Mould, magnified.

rudiment of a stem exists underground in what is called a *rhizome*, from which the *fronds* shoot out, in the same manner as the leaves spring from the buds of other plants; these fronds have a strong midrib which is commonly called a stalk. There are said to be between two and three thousand varieties of Ferns; some of them, in the tropics, attain the enormous height of thirty feet. Their growth is extremely interesting, the fronds opening in a peculiar manner, unwinding themselves, as it were, from a round ball. The seed-vessels are placed on the back of the fronds in little spots or bunches, and the seed is so fine as to be only perceptible under the microscope. Ferns thrive best in moist and warm situations; if grown under a glass vessel which will confine the moisture, they form a beautiful and interesting parlor ornament.

Ferns.

CHAPTER VI.

ARCTIC PLANTS—VARIETIES OF CLIMATE AND EFFECT UPON VEGETATION—RHODODENDRONS—TEA—MODE OF PREPARATION — BARREN PINE — PITCHER PLANT — SPIKENARD — SAFFRON — CROCUSSES — MOTION IN PLANTS — SENSITIVE PLANT — VENUS FLY-TRAP — ROOTS OF PLANTS.

T is very interesting and instructive to examine into the character of the different plants which are adapted to various sections of the globe.

While there is but little doubt that Nature no where displays her gaudy colorings in greater profusion, or to better advantage than in the wilds of South America, yet there are many other lands where the productions of the vegetable kingdom are no less useful and attractive. Even the ice-bound regions of the Arctic Circle can boast of their green mossy banks and smiling flowers, which are certainly none the less remarkable for the fact that, owing to the shortness of the summer season, the process of vegetation is so rapid, that in some species the whole time required to reach maturity is little more than a month.

ALPINE PLANTS.

Many Alpine plants, cradled in perpetual snows, and exposed during a great part of the year to the driving of the wintry blasts, which are so common in Switzerland, Lapland, and other cold regions, are so tenacious of their accustomed haunts and habits, that

> "The raging tempest and the mountain's roar,
> But bind them to their native hills the more;

and any attempt to grow them in a milder climate is generally attended with failure. These plants are mostly quite diminutive, although they sometimes produce flowers of considerable size and beauty.

The most common color among plants which inhabit very cold countries is white, or a light shade of pink or yellow. Thus, the snow-drop, the lily of the valley, the white-flowered wood-sorrel, are all productions of high northern latitudes; while in warmer regions, the flowers are robed in stronger hues.

It is observed that mountainous places are generally much more productive than the valleys; but there is scarcely any situation, however unfavorably located, where plants and flowers are not occasionally met with. They are found

> "Springing in valleys green and low,
> And on the mountains high;
> And in the silent wilderness,
> Where no one passes by."

On one of the highest points in Europe, at the elevation of eight thousand feet above the level of the sea, is a beautiful and verdant garden, which is entirely surrounded by snows that never melt. This spot is covered with Alpine plants; and so luxuriant is the growth of the vegetation, that at certain seasons of the year the Swiss peasants drive their cattle over the great glacier of Mer de Glace for the sake of the delightful pasture the valley affords.

In our own country, where so great a variety of climate is witnessed, it is probable that a greater variety of plants can be enumerated than in any other. Our gardens and conservatories are indebted for many of their finest ornaments to the far-off fields and woods of California, Mexico, and the territories west of the Rocky Mountains,—countries which combine within their range a climate varying almost from frigid to tropical, and exhibiting at the same season a corresponding difference in their floral productions. In the northern and western States, while the cold earth still lies locked in winter's last embraces, the woods of the south are teeming with life, the fields are clothed with the verdure of spring, and the air is scented with the perfume of flowers. But in the regions of tropical Mexico, and the everglades of Florida, vegetation becomes so entirely changed in its character, as to maintain a more uniform appearance at all seasons of the year.

Most plants, whose roots are perennial, have a period of rest, during which they cease to grow; in the north, this is usually indicated by the falling of the leaves, and the plant assuming the appearance of being dead; in the tropics it is marked by the absence of flowers, and of the fresh and vivid green of the younger growth.

Rhododendron.

But in many plants, even in rigorous climates, this period of rest is not attended by the falling of the foliage. The beautiful varieties of the Rhododendron, some of which inhabit the mountains of Pennsylvania, are examples, among many others, of evergreen shrubs. The greatest variety of these superb flowering plants grow on the woody slopes of the Himalaya Mountains, where they may be seen early in the spring loaded with their conspicuous heads of often gay-colored and fragrant blossoms. Occasionally large trees become quite embowered in them, as they sometimes fasten themselves to the trunks, and, leaving their hold upon the earth, creep to the very summits, where they grow in the manner of parasites, deriving their nourishment from the bark.

The flowers of these plants vary much, both in size and color; some are very large, and appear two or three together; these are mostly white or cream-colored, resembling a lily; others are brilliant crimson, deep scarlet, rose-colored, yellow, or purple, and hang in large bunches at the ends of the branches. Travellers in the Himalaya Mountains speak of the Rhododendrons as being among the most beautiful of the many vegetable curiosities of that fertile region.

There are also many plants whose leaves, as well as their flowers, form objects of wonder; and some are rendered peculiarly interesting because of the prominent part they occupy in our domestic economy. Thus, the common Tea Plant is so well known, that every one should be made acquainted with the mode of its culture, as well as with the method of converting the leaves into that useful article, which takes so conspicuous a place in the commerce of the world.

This plant grows about eight feet in height, with leaves two and a half inches long, and one and a half wide, and bears a small white flower. The Chinese raise the plants from seed, and when they have grown of sufficient size, they are set out in the ground at intervals of about three or four feet apart; they are kept cropped close for a year or two, to make them grow thick and bushy. When they are about four years of age, they commence to gather the leaves; this is done several times during the year, and is continued for about six or eight years, when they are removed and fresh ones planted. The leaves first gathered in the spring make the finest flavored teas, while those which are taken subsequently produce a much inferior article.

What are commonly known as green and black

teas are the products of the same plant, treated in different ways. The green tea is made by commencing to dry the leaves in the ovens as soon as they are picked, the whole operation of drying, rolling, and roasting, being done very quickly; while in the black tea the leaves, when picked, are laid in the sun until they become entirely soft and wilted, when they are shaken about in sieves held over hot steam; this deprives them of the peculiar properties which belong to the green tea. When the leaves become quite flaccid and watery, they are put into large copper dishes and roasted for a few minutes over a hot fire, when they are taken out and rolled between the hands. In the finer sorts, each leaf is rolled separately; after this, the process of drying and baking is commenced by alternately placing them over the fire, and then exposing them to the air for some hours. This is repeated five or six times, when the tea is fit for use.

Tea leaves possess properties which will produce giddiness, headache, and even paralysis; these properties are much weakened in the process of drying; and the longer this is in being completed, the more wholesome tea becomes. Both green and black teas act as powerful nervous stimulants upon a system which has not become accustomed to them; hence the benefit often derived from their use by persons in advanced life who have abstained from them when young.

The tea plant was cultivated, and its leaves used, as early as the fourth century; and, in the year 763,

Tea Plant.

a duty of ten per cent. was laid upon it by the Chinese government; since which time it has been a fruitful source of revenue to the Empire. The annual product of China alone amounts to the enormous quantity of two and one-half billions of pounds — (2,500,000,000.) Add to this the vast product of Japan, Java, and Corea, and we may justly be amazed to think what a great tea-drinker the world is.

Wherever we turn our eyes, and from whatever point we view the vegetable kingdom, we see new wonders; something new to be learned; and as we are always forgetting, how well it is that new subjects of interest are always awaiting our notice. And how instructive it is in all these things to observe that Nature adapts herself to the peculiar circumstances in which she is placed.

The barren pine, so called from its being unproductive, exactly resembles the stem that bears the pine-apple in our green-houses. It is not, however, entirely useless; for in some species there is a protuberance hanging down resembling a bowl; in this the rain collects, and remains a considerable length of time quite pure and sweet. This, Nature provides for the use of the plant. It grows on the dry stump of a withered tree, and from the sapless wood it could derive no nourishment; and thus a new mode of supplying it with moisture is found. Nor is this all; the plant generally grows on trees on the tops of mountains, where there are neither streams nor springs, and in hot weather it frequently yields the traveller a cool and refreshing draught, when no other water can be found near it.

There are also some plants which spring up in dry and sunburnt soils, whose herbage is of so juicy a nature, as to serve the same purpose as water in quenching the thirst. But perhaps the most remarkable plant yet known, which possesses the faculty of secreting pure water, is the Pitcher Plant.

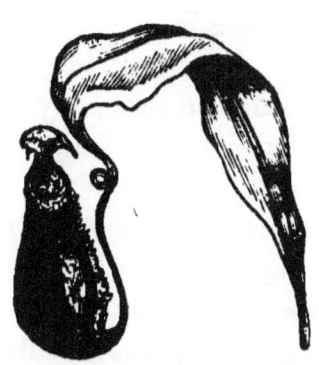

Pitcher Plant.

From the end of each leaf hangs a large vessel in the shape of a pitcher, and capable of holding nearly a pint of water; each pitcher has a lid fitting closely to the top, and opening wide upon its hinges in damp weather, and again closing when it is dry, to prevent evaporation. But how, it may be asked, is this delicately suspended vessel supported when so full? Nature here supplies an adequate provision; behind the lid is placed a little hook, which, with marvellous sagacity, catches hold upon some neighboring twig or tendril, and thus the required support is obtained. The liquid is secreted by the plant itself, and is remarkably pure, though it grows in a muddy and unwholesome marsh.

The costly and delicious odor, known in ancient times as Spikenard, has been assigned by various authors as the product of different plants, and it has not been until recently that any satisfactory information has been gained respecting it. There now appears, however, to be but little doubt that it is a species of Valerian, which grows in the colder parts of the mountains of India. The plant must have been comparatively rare, as it is always spoken of in Scripture as being very precious or costly. When Mary anointed the head of Jesus with this sweet perfume, Judas took occasion to murmur at the waste, saying, that "this ointment might have been sold for more than three hundred pence, and given to the poor." It was usually imported in boxes of alabaster, and when the master of a house received his guests, it was customary for him not only to crown them with flowers, but also to bring forth the box of precious ointment, and break the seal which prevented the volatile perfume from escaping, and anoint them with it. So sweet was its fragrance, and so much esteemed, that Horace, speaking of it, says, "that a small onyx box full was equal in value to a large vessel of wine." Its costliness was probably owing in part to the great distance from which it was brought.

In reading over the books of the Old Testament, one cannot fail to be struck with the fact that perfumes were held in great esteem by the Hebrews, and that very large quantities were used by them.

"Spikenard and Saffron, Calamus and Cinnamon, with all trees of frankincense, myrrh and aloes, with all the chief spices," (Cant. iv. 14,) seem to have been very prominently useful in the preparation of odors which were then esteemed so important in the performance of many religious rites. It was among these, as above enumerated by Solomon, that we have the first mention of Saffron.

There are few, perhaps, who are not well acquainted with the common Crocus of our gardens, which in the spring, in company with the Snowdrop, fairly pierces the snow and ice, to gain admission to the light, and expand its beautiful blossoms. The Saffron, although it blooms in the fall instead of the spring, is nevertheless a genuine crocus. In some parts of England it is extensively cultivated, and has almost become naturalized, and may often be seen spreading out its bright purple flowers to the sun, in the meadows and pasture-fields.

The article, known as Saffron in commerce, is nothing more than the dried stigmas taken from the flower; they are of a very brilliant yellow color, and possess an agreeable odor.

The power of motion, similar to that of animals, is often seen in the vegetable kingdom. It is very common for climbing plants to bend their stems toward some object upon which they may obtain support; this can hardly be called motion, as the change of position is so gradual as scarcely to be perceptible.

F

But in some of the species of Mimosa or Sensitive Plants, the slightest touch of the hand will produce a sudden and very considerable change in the position of the leaves, as they will immediately fold themselves together, and if the touch be repeated, the leaf-stalks will fall and bend themselves toward the stem. A strong wind or heavy rain will produce the same effect, and those kinds which grow in countries where there is a long continuance of rain, close their leaves upon its approach, and are seldom fully expanded until the return of fair weather.

The leaves of the common sensitive plant will partially close at evening, and remain so until the light of day again causes them to expand; and when exposed to sudden cold during the daytime, they will fold themselves face to face, so as to allow as little as possible of the upper surface to remain open to the air.

The Venus Fly-trap is also an example of motion; the leaves are very curiously constructed. At the extremities are spread out two wide and rounded lobes, each armed upon the margin with rows of spines, or rather stiff hair-like processes, locking into each other when they meet, which always occurs when irritated. The upper surface of these lobes is covered with minute glands, which evidently contain a liquid attractive to insects, as they frequently resort to the plants; but no sooner do they alight upon the leaves, and their little feet irritate them, than the two lobes instantly fold together, squeezing the insect to death. The leaf seldom opens again, unless the prisoner is first set free; and so powerful is the grasp by which it is held, that it requires considerable force to liberate it. The plant cannot, however, discriminate between the touch of a straw and the tickling of a bee, as the effect of either is the same.

The peculiarity of the roots of different plants is as noticeable as that of the leaves and flowers; and although they possess nothing that is attractive to the eye, they are nevertheless essential to the existence of the plant.

They have been, for the most part, placed by the Creator below the surface of the ground, yet they are not to be entirely lost sight of on this account. Let us learn from them not to despise those whose circumstances in life are apparently beneath our own, but ever remember that the proud and majestic oak, that waves its tall summit in the breeze, unconscious of the root that bears it, is none the less dependent on that root for its safety in the storm.

All plants have either annual, biennial, or perennial roots. Annuals are such as continue but one year, the plant reproducing its species by seed; biennials are such as spring from the seed, forming a plant during the first year which does not mature its seed until the second season; and perennials are such as live for an unlimited length of time, making fresh growth and producing seed each year.

The most common form of roots is fibrous; these are divided and subdivided into minute filaments which often penetrate the soil to a great distance. It has been said by some authors that the roots of trees spread as much beneath the ground as the branches do above it; but this can hardly be said to hold good in all cases. Fibrous-rooted plants often perform great service in loose sandy soils, especially along water-courses, where they form a thick and matted mass, thus preventing the washing away of the earth. Tuberous roots are solid and very irregular in their shape, and are often linked together by slender fibres. Roots of this form are the most useful, as they are generally edible; the common potato, the turnip, and the radish, are familiar examples.

Bulbous roots are of various kinds; some are solid, as in the crocus; others are composed of fleshy layers placed one above the other, as in the onion; and others consist of thin scales, as in some species of the lily. They all appear to act as reservoirs for the vitality of the plant during its dormant state.

A bulb is entirely analogous to the bud upon a tree, each containing within itself the embryo of the future stem or plant. In the bulb of the tulip, the microscope will reveal the entire leaf, stem and flower, all folded up within its layers, and which require nothing but the action of light, heat and moisture, to expand into perfection; so in the bud upon the tree, the leaves and blossoms which open in the spring are all encased in miniature in that tiny compass.

CHAPTER VII.

TREES.

USES OF TREES—THE OAK—VARIETIES OF THE OAK—STRUCTURE OF THE ACORN—GROWTH OF THE TREE—CIRCULATION OF THE SAP—EFFECT OF LIGHT UPON THE FOLIAGE—LIVE-OAK — CORK-OAK — HOLLY-OAK — "OAKS OF BASHAN" — BLACK AND WHITE-OAKS — STATISTICS OF THE SIZE AND AGES OF VARIOUS OAKS IN AMERICA AND EUROPE — THE CHESTNUT AND THE CHESTNUT-OAK—AGE AND SIZE OF THE CHESTNUT—THE BEECH—THE WALNUT—THE HICKORY.

LET us now turn our attention from the beautiful verdure which clothes the surface of the earth, and behold the wonders of creative skill, as displayed in the majestic Trees of the Forest. How gracefully they bend their waving summits to the passing gale! How softly murmurs the fragrant breath of summer through their leafy bowers! How gorgeous are the tints in which sere autumn robes them! While dark and dreary winter, with its thousand storms, wraps its grey mantle around their naked branches.

How indispensably necessary to our comfort are trees! How endless are the uses to which they are applied! To say nothing of their beauty and the charming freshness of their shade, how multiplied are the conveniences which they afford by supplying us with that most useful article, Wood!

Without trees for building purposes, and for the manufacture of those many utensils, which seem almost coupled with our very existence, how changed would life be! Half the comfort of our houses is due to the wood which forms so large a part of the materials of which they are constructed; and many of the charms of country firesides are owing to the unconscious and unsightly logs that lie blazing upon the hearthstone.

But however dependent upon trees for the supply of our daily wants, and however they may excite our admiration when we look upon their noble forms, yet how few there are who sufficiently reflect upon the manifold blessings which are conferred upon us by their existence!

The forest trees of the temperate zone may be considered as forming the type of the second great natural order of plants, called "Exogenous," from the fact that the new growth takes place on the exterior surface of the stem, a new layer of wood be-

Exogenous Wood, showing the growth of nine years.

ing deposited each year. The leaves of such plants have their veins running in all directions, forming a fine network.

It is to this class of trees that we principally look for our supply of timber for building and other purposes, as they are far more abundant than the endogenous, and attain to a much greater size, while the wood is more easily worked, and much more durable.

The most useful woods in supplying our daily wants, both as fuel and building-lumber, are Oak, Chestnut, Pine, Hemlock, Elm, Ash, Hickory, Poplar, and Maple.

Those which are most useful as fancy-woods for the manufacture of Cabinet-ware, are Mahogany, Walnut, Rose-wood, (obtained from a species of Mimosa which grows in Brazil,) Tulip-wood, (to be had only in small pieces not wider than five inches,) Zebra-wood, (probably the production of a large tree, and beautifully shaded with white, red, and black stripes,) Satin-wood, (a fine-grained wood of a brilliant yellow color, brought from India,) Sandal-wood, (resembling tulip-wood, and possessed of a very fine odor,) Camphor-wood, (the product of the Camphor-tree,) Ebony, Iron-wood, Canary-wood, and many others of less importance.

The wood of the Box-tree is also an article of considerable importance in commerce; it is remarkably fine and close-grained, which makes it particularly serviceable for the use of the engraver; and it is to this tree that we owe the facility and cheapness with which many books are illustrated.

The variety to be observed in trees is not so great as in other plants; but in the same forest a considerable number of kinds may be seen that are entirely different, even of such as are commonly known by the same name. Thus, what is termed Oak, often consists of eight or ten varieties, such as Black-oak, White-oak, Chestnut-oak, Pin-oak, Willow-oak, Red-oak, Scarlet-oak, Spanish-oak, Post-oak, &c. &c. These all differ, not only in the formation of their leaves and fruit, but there is a marked difference in their manner of growth, and the wood of each possesses its own peculiar properties. Within the limits of the United States, there are no less than thirty or forty varieties of this useful tree; some are quite small, growing only to the height of two or three feet; but by far the largest number are lofty trees, with wide-spreading branches. Let us here stop and pick up an acorn which has fallen from one of these, and examine its wonderful construction, and compare its tiny proportions with the majestic plant that bore it, and the counterpart of which it is destined to produce. Encased within that bony covering lie hid all the essential parts of the infant tree, perfect and complete; roots, stem, leaves and buds.

White-Oak.

The germ of the future plant is placed at one end of the acorn, and although of so delicate and fragile a texture, that a slight rub would be sufficient to break it, yet so nicely is it fitted to its shell, that the nut may be handled very roughly without injuring it. This germ consists of two parts, the plume which rises and forms the future stem, and the beak or radicle which descends and forms the root. How surprising is the ascent of the one and the descent of the other! It is, in fact, the

Black-Oak.

effort of the one to get into the air, and of the other to enter the earth. Were they to be placed in an inverted position, the result would be the same; each would bend itself toward its proper element.

Clasping the germ are the two lobes of the kernel, which serve the important purpose of sustaining the life of the plant until it has become sufficiently rooted to derive all its nourishment from the soil.

When through the combined influence of heat and moisture, the germ cracks open the case by which it is confined, it sends down a strong radicle, called a

Red-Oak.

tap-root; then the two lobes of the kernel separate, and the plume springs out from between them. This consists of two leaves, which soon expand and disclose

8*

at their base a bud from which in like manner additional leaves make their appearance. The functions of the little plant are now as complete as they are in the monster tree. The delicate stem is supplied with its capillary tubes, which carry the nourishment from the root, and distribute it throughout the different

Post-Oak.

parts of the plant. These vessels perform the same part to the tree as the veins and arteries to the human body. One set, comparable to the veins, carry the sap through the trunk and branches to the leaves, where it is spread out in the minute reticulated net-work on their upper surfaces, and like the blood in the lungs is there exposed to the action of the light and air, absorbing from the latter a portion of carbon, and at the same time parting with its oxygen. This process purifies the sap, and renders it fit for the nourishment of the tree, when a new set of vessels, similar in their office to the arteries, distribute it to the different parts where it is required. This sap, thus purified, contains all the requisites for the formation of the wood and bark of the trunk and branches, and the cuticle for the formation of the leaves.

The effect of light upon the sap is very great; those plants which receive the largest amount of sunlight have leaves of a deep green. Light is therefore an essential element in promoting the healthy growth of trees, as it will be observed that the absorption of carbon, and the giving out of the oxygen becomes less active as the light is diminished, while during the night the contrary process occurs, the oxygen being absorbed, and the carbon released.

Thus it is by the assistance of these vessels that the little oak-plant becomes a perfect tree in miniature, and continues to increase in size and strength from year to year, until the woodman levels with his axe the sturdy trunk that has defied the storms of a century. Let us here examine the stump that remains, and we will observe a number of concentric rings commencing at the bark, and running around the tree, one inside of the other, until they reach the centre. (See figure of Exogenous Wood on a previous page.) Each ring indicates one year's growth, a new layer of wood being deposited every season immediately under the bark. The age of any tree may be determined with considerable accuracy by counting these rings.

Chestnut-Oak.

Some species of Oaks retain their foliage during winter, as the Live-oak. This tree inhabits the Southern States, where it occasionally grows to

Spanish Oak

a considerable size. It is probably the most valuable wood known for ship-building, on account of its great durability. In South Carolina the Live-oaks are often hung with the graceful festoons of a beautiful moss, which dangles from their branches in pendent masses of several yards in length.

The Cork-oak, a native of the South of Europe and the northern coast of Africa, is also an evergreen, and much resembles the Live-oak in appearance. That useful substance, Cork, is the bark of this tree, which grows to a considerable thickness; and, as though designed by Providence for some peculiar purpose, may be removed without injury to the tree, a new coating being rapidly formed; thus producing a crop of cork about every ten years.

Live-Oak.

The Oaks of Palestine are also mostly evergreen. One species, closely resembling the Holly in its appearance, and called the Holly-leaved Oak, is particularly abundant; the scenery of the Holy Land being often varied with its beautiful form.

As the climate of Syria is too warm for the Oak to flourish in the valleys, it is mostly confined to the more elevated positions. Groups of low shrubby Oaks are scattered all over the hills of Hebron; and many of the evergreen varieties are found in the forests which cover the hills of Canaan. But the tall "Oaks of Bashan," spoken of in Scripture, are more attractive on account of their great size and venerable appearance, as well as their luxuriant foliage; and many a weary traveller may repose beneath their refreshing shade upon the spot rendered memorable by the beautiful similes of the Prophets Isaiah and Zechariah. Isa. ii. 12, 13. Zech. xi. 2.

Of all the varieties of the Oak, the black and the white are with us the most abundant and the most useful; the wood is extensively used in ship-building, as well as for many other purposes equally important, while the bark is not only the principal substance used by the tanner in the preparation of leather, but is also of great use in dyeing.

Willow-Oak.

These Oaks often attain a great size, and live to a very advanced age. The "Charter-Oak," so celebrated in the history of our country, was an old and venerated tree in the Revolution. The "Flushing-Oaks," the remains of which are still standing, yielded an abundant shade, under which large congregations were accustomed to assemble near two hundred years ago to listen to the preaching of George Fox.

In England, where antiquity is more venerated than in America, such relics of bygone ages, sentinels that have watched over the destiny of many a monarch, are regarded with deep interest. The ages of some of these have been computed with considerable certainty, by reference to data which have been preserved on record relating to them, and also by means of some inscriptions which have been found deeply imbedded in the solid wood, and over which the growth of years has been deposited.

The following description of a few remarkable trees in different parts of England, is taken from Loudon's Arboretum:—

"The Merton Oak stands on the estate of Lord Walsingham. It is 66 feet high, and, at the surface of the ground, the circumference of the trunk is 63 feet 2 inches. At one foot [from the ground], it is 46 feet 1 inch; the trunk is 18 feet 6 inches to the fork of the branches; the largest limb is 18 feet, and the second 16 feet in circumference."

"The Winfarthing Oak is 70 feet in circumference; the trunk is quite hollow, and the cavity large enough to hold 30 persons. It is said to have been called the "Old Oak," at the time of William the Conqueror.* It is now a mere shell—a mighty ruin, bleached to a snowy white; but it is magnificent in its decay. The only mark of vitality which it exhibits, is on the south side, where a narrow strip of bark sends forth a few branches, which even now (1836) occasionally produce acorns."

"The Salcey Forest Oak is described as 'one of the most picturesque sylvan ruins that can be met with anywhere.' It is supposed to be above 1500 years old; and its trunk is so decayed as to form a complete arch, which is 14 feet 8 inches high, and 29

---

* This tree is most probably 1500 years old.

feet in circumference inside. The tree is 33 feet 3 inches high, and 47 feet in circumference on the outside near the ground. This fine ruin is still standing, and, though it has latterly become much wasted, it annually produces a crop of leaves and acorns."

The Chandos Oak.

"The 'Chandos Oak' stands in the pleasure-grounds of Michendon House, near Southgate, and is about 60 feet high. The head covers a space, the diameter of which measures 118 feet. The girth of the trunk, at one foot from the ground, is 18 feet 3 inches. It has no large limbs; but, when in full foliage, its boughs bending to the earth, with almost artificial regularity of form, and equi-distant from each other, give it the appearance of a gigantic tent. It forms, indeed, a magnificent living canopy, impervious to the day."

"The 'Boddington Oak' grew in a piece of rich grass-land, called the Old Orchard Ground, belonging to Boddington Manor Farm, in the vale of Gloucester. The sides of the trunk were more upright than those of large trees generally; and at the surface of the ground it measured 54 feet in circumference. In 1783, its trunk was formed into a room which was wainscoted, and measured in one direction 16 feet in diameter. The hollowness, however, contracts upwards, and forms itself into a natural dome. It is still perfectly alive and fruitful, having this year (1783) a fine crop of acorns upon it. This tree was burnt down, either by accident or design, in 1790."

"Of the Magdalen, or Great Oak of Oxford, Gilpin gives the following interesting notice:—'Close by the gate of the water-walk of Magdalen College, Oxford, grew an Oak, which, perhaps, stood there a sapling when Alfred the Great founded the University. It is a difficult matter to ascertain the age of a tree. The age of a castle or abbey is the object of history. But the time occupied in completing its growth is not worth recording in the early part of a tree's existence. It is then only a common tree; and afterwards, when it is become remarkable for age, all memory of its youth is lost. This tree, however, can almost produce historical evidence for the age it boasts. About 500 years after the time of Alfred, William of Waynfleet, Dr. Stuckely tells us, expressly ordered his college [Magdalen College] to be founded near the Great Oak; and an oak could not, I think, be less than 500 years of age to merit that title, together with the honor of fixing the site of a college. When the magnificence of Cardinal Wolsey erected that handsome tower which is so ornamental to the whole building, this tree might probably be in the meridian of its glory; or rather, perhaps, it had attained a green old age. It was afterward much injured in the reign of Charles II., when the present walks were laid out. Its roots were disturbed; and from that period it declined fast, and became reduced to a mere trunk. Through a space of 16 yards on every side from its trunk, it once flung its boughs, and under its magnificent pavilion could have sheltered with ease 3000 men. In the summer of 1788, this magnificent ruin fell to the ground."

"The Cowthorpe Oak, in Yorkshire, measures at its base 78 feet in circumference. The space occupied by this tree, where the trunk meets the ground, exceeds the ground-plot of that majestic column, the Eddystone Light-house; and horizontal slices of Damorey's Oak would have laid every floor in one piece throughout the whole building."

The oak and the chestnut are very closely connected, not only in their appearance, but also in their general character. The leaves of the chestnut and the chestnut-oak would be mistaken for each other by one unaccustomed to the difference, those of the chestnut being only a little more sharply toothed than the other.

In California, a species of oak has been discovered, whose mode of flowering, and indeed the whole appearance of the tree, is so similar to that of the chestnut, as to require the presence of the fruit fully to determine its identity. The wood of each also bears some comparison; the color and grain being much the same; the oak is, however, tougher and heavier, while the chestnut, in consequence of the evenness and regularity of the fibre, possesses the peculiar property of being easily split into long straight pieces. Hence its great utility to the farmer for fencing.

A close connection is also observable in many of their habits. The chestnut delights in a high and hilly soil, and grows freely in the same positions where the chestnut-oak abounds. If a large tree of each be cut down, strong scions will soon spring up in all directions from the roots, forming bushy clumps, which resemble each other so closely as to be readily taken for the same tree.

The chestnut also lives to a great age, and in some situations grows to an enormous size. The famous chestnut tree, which grew upon Mount Etna, was probably one of the largest and oldest trees in the world. In 1770, this tree is said to have measured 204 feet in circumference; its trunk was quite hollow, and a house had been built in the interior, which was inhabited by some country people. The age of this tree of course cannot be estimated with any certainty.

The old chestnut tree at Tortworth, in England, was probably planted by the Romans, as the tree is not a native of that country. It was evidently old at the time of the Norman Conquest, as history speaks of it as a famous tree in the time of King John. It measured 57 feet in circumference.

The Beech, in some respects, resembles both the Oak and the Chestnut; and was originally classed with the latter by Linnæus, the great Swedish Naturalist. The wood, however, differs much from the others in being very close and fine-grained. The fruit is enclosed in a scaly burr, somewhat resembling the cup of an acorn, which, when matured, opens into four sections, and allows the triangular nuts to escape. In France and Germany, an excellent oil is obtained from the kernels, which is said to be superior to that produced by the Olive.

The beech is rarely found living to any great age, although occasional specimens are met with which are evidently of great antiquity. A Beech which stood some years since in Windsor Forest, England, is said to have existed prior to the Norman Conquest, which would indicate that it had known the changes of at least 800 years. At the time of the last measurement, it was about 36 feet in circumference at the base.

In America the beech is a beautiful tree, with dense and finely-cut foliage, forming a thick and impenetrable shade. It sometimes attains a height of 100 feet, with a trunk measuring 8 or 10 feet in circumference.

In connection with the Chestnut and Beech must be mentioned the Walnut and Hickory; trees of great beauty and interest, as well as utility. Of each, there are several varieties. Of the Walnut, the Black is probably the most useful by far, it being used very extensively in this its native country, as well as in Europe, for the manufacture of cabinet-ware. The wood, which is of a fine dark color, and beautifully veined and mottled, is susceptible of a very high polish. Some of our most beautiful articles of furniture are made from this wood, and it may be justly ranked among the most useful of our sylvan productions. The black walnut occasionally, though seldom, attains a great size. The trunk of one grown on the south side of Lake Erie, was some years since exhibited in London, which was 12 feet in diameter, and was hollowed out and furnished as a sitting-room. The tree was said to have been 150 feet in height, with branches from 2 to 4 feet in diameter, and the bark 1 foot in thickness.

The Hickory, though nearly allied to the Walnut, possesses properties peculiarly its own; its wood is light-colored, tough, and elastic, which renders it very serviceable to the carriage and wagon builder; and the air of comfort which always surrounds the hearth where the crackling of a good hickory fire is heard, fully attests its usefulness as fuel.

The Hickory, particularly the variety known as the Shellbark, is a noble and majestic tree, rising to the height of 70 or 80 feet, with a trunk sometimes 5 feet in thickness at the base, and varying but little from the straight line almost to its summit, and frequently without a branch below the height of 40 feet. The gathering of the nuts of the walnut and hickory affords considerable merriment to the younger part of the farmer's family, while many a city fire-side, cheered by the social gathering, has found a rich treat in the fruits of these noble trees.

CHAPTER VIII

THE WILLOW — NUMBER OF VARIETIES — NAPOLEON'S WILLOW — CURLED WILLOW — ELM — BIRCH — POPLAR — ASPEN — LOMBARDY POPLAR — TULIP TREE — THE YEW — FLOWERING TREES — THE MAGNOLIA.

"BY the rivers of Babylon there we sat down; yea, we wept when we remembered Zion. We hanged our harps upon the willows in the midst thereof." Ps. cxxxvii. 1, 2.

This beautiful and poetic allusion, undoubtedly refers to the Weeping Willow, which was formerly very abundant in the environs of Babylon, whence arises its botanical name, Salix Babylonica. The word Salix is derived from the Celtic, and means near water, referring to the general habit of all the willows of frequenting watery places. They often give a very picturesque appearance to the landscape, as they spread their branches, covered with the most beautiful foliage, over the smooth surface of the water, or gracefully dip their long slender boughs into the stream.

But we may imagine that, however beautiful was the effect thus produced, it must have possessed but few charms for the captives of Judea, as they sat mournfully brooding over their sorrows, with harps unstrung, and weeping at the remembrance of Zion's surpassing loveliness.

Many of the common varieties of willow are perhaps known to most of our readers. But there are few who are aware of the great difference that really exists between many that appear to be the same. There are probably no less than forty or fifty distinct varieties to be found in the United States, and more than double that number in other parts of the world.

In the Arctic regions there is a species which is no more than a few inches in height; and in latitudes nearly approaching the pole, it is almost the only woody plant to be found. The Weeping Willow grows in China, Japan, Syria, and the northern parts of Africa, which appear to be its native localities; but it may also be seen in most of the countries of the temperate zone.

In the island of St. Helena there once stood a tree of this kind, which was known as Napoleon's Willow. It was planted by the Governor of the island about the year 1810, and grew among the other trees on the side of a valley, near to a spring. Having attained a considerable size, it attracted the attention of Napoleon, who had a seat placed under it, and used frequently to resort to its shade, and have water brought to him from the adjoining fountain. About the time of the death of the Emperor, it is said that a storm shattered the tree in pieces. Many cuttings were taken from it; and trees propagated from this original may now be found in various parts of the world. By many, this tree was supposed to have been of the variety known as the Curled Willow; but this appears to be an error.

The Curled Willow, whose leaves are curled into rings, or twisted up like corkscrews, is nothing more than a curious variety of the Weeping Willow; it is of rather a dwarf habit, and the crisp and parched appearance of the leaves destroys much of the beautiful effect of the drooping of the branches

Scarcely anything, it may be said, enters so deeply into the beauty of a landscape as the great variety noticeable in the outline presented by different trees, as well as the multiplicity of the shape, size and color of the foliage. The tall spire of the Lombardy Poplar, with its small, opaque leaves, peers far above the rounded tops of the Maples and Lindens; and the sharp-pointed cone of the Cypress forms a fine contrast with the irregular outline of the Tulip Tree; while on the deep, dark back-ground, formed by the large and heavy leaves of the Oak and Hickory, stands out in pleasing prominence the fine, light, and silvery foliage of the Willow.

The Elm also assumes a very prominent position in the American Landscape; and the eye cannot fail to rest with pleasure upon its beautiful outline. It is in the northern and eastern States that it attains the greatest perfection. The trunk rises to the height of 60 or 70 feet, insensibly diminishing in thickness from the base, until it is lost in the minute ramifications of the topmost boughs, which are widely divergent, and shoot out on all sides in long, flexible, and pendulous branches, bending into regular arches, and floating lightly in the air. In isolated positions the Elm occasionally grows to the height of 100 feet; the trunk is then sometimes clothed to near its base with its beautiful verdure, which seems to wreathe about it like some parasitical vine or creeper.

In Europe, the Elm lives to an advanced age, and often attains a prodigious size. The Crawley Elm, situated on the road from London to Brighton, is 71 feet high, and the trunk measures at the ground 61 feet in circumference.

At Hampstead, a Hollow Elm formerly stood, the trunk of which measured at the base about 30 feet, and at the height of 42 feet appears to have been broken off. It is entirely hollow from the top to the bottom, in which a staircase had been built, leading to the summit, which was turreted, and provided with seats for six persons. It appeared to be in a thriving condition, and covered with the most luxuriant foliage, which spread to a considerable distance on every side.

The Birch and the Poplar must also be reckoned among the beautiful trees of our forests. The Paper, or Canoe Birch, and the Yellow Birch, both inhabitants of the Northern and Eastern States and Nova Scotia, are lofty trees, with their trunks measuring from 3 to 18 feet in circumference. The former is remarkable for the beautiful texture of the bark, which is capable of being split into thin plates or layers, which have a fine smooth surface, and when carefully prepared, may be used as a substitute for paper. The thicker plates are made into canoes by the Indians, which are particularly light and buoyant, and entirely impervious to water. One of these, constructed to accommodate four persons with their baggage, it is said, will weigh only 40 or 50 pounds. Their lightness renders them peculiarly serviceable in navigating rivers where the stream is often interrupted by rocky rapids or cascades, as they may be readily carried around them by land, and again launched in the water below.

Aspen.

In the settlements of the Hudson's Bay Company, we are told this tree sometimes measures 18 or 20 feet in circumference at the base; the bark is used in building tents, it being cut in pieces 12 feet long, and 4 feet wide; these are sewed together with the long pliable roots of the Spruce, and so rapidly is the work done, that, a tent of 20 feet in diameter, and 10 feet high, does not, it is said, occupy more than half an hour in pitching.

No small quantity of Birch-wood was used by the School-masters of the Olden-time, as a means of instilling sound views and correct principles into the minds of their pupils; but thanks to the progress of civilization, that practice is almost obsolete, and the once-dreaded birch is again consigned to those uses for which it was originally created.

Among the varieties of our native Poplars, perhaps the most beautiful and the most familiar is the Aspen. The leaves of this tree are hung on long and slender petioles, which are flattened at the base, and attached to the stem in such a manner, as to cause them to rock backward and forward, with the least motion of the air. When scarcely the slightest breeze is stirring, and every other tree seems lulled into complete repose, the foliage of the Aspen may often be seen quivering as though shaken by force.

The Lombardy Poplar, once so abundant about our farm and country houses, and which shoots above its surroundings, like some tall church-spire, was originally brought from Italy, where it abounds on the banks of the river Po. Its cultivation is now becoming much neglected, partly in consequence of the climate or soil not being adapted to its growth, as its beautiful and pleasing outlines are often marred in a single season by parts of the tree being killed, either by the severe cold or other causes. It is much to be regretted that this is the case, as it certainly forms a very conspicuous object, and occupies a position in our rural scenery which no other tree can supply.

In Europe, it attains to a great size and beauty, often measuring over 100 feet in height, and 3 to 5 feet in diameter at the base.

The tree, commonly known as the Tulip Poplar, belongs to a very different genera, and is altogether misnamed, as it bears no resemblance whatever to the Poplars. It is, however, one of the most useful, as well as the most beautiful, of our forest trees. The wood is commonly called poplar-wood, and being soft and easily worked, is extensively used in the manufacture of Cabinet-ware. This tree is very conspicuous in the early summer months on account of the abundance of its large showy flowers, each being the size, and having much the appearance of the tulip. But its appearance is too familiar to need much further description.

A noble specimen of this tree, which recently stood upon the farm of Friends' Boarding-school at West-town, measured at the base about 37 feet in circumference, and was about 100 feet in height. It was hollow in the centre, with an opening on one side like a tent-door. Respecting its age and history, one of the Principals in the Seminary writes:—" We have no *data* from which to determine its age, but judging from analogy, it must have been in existence *long* before William Penn founded the colony. The importance with which this tree was regarded was no doubt mainly due to a tradition that it was once occupied as a dwelling by a family of Indians. The tradition most likely had its origin in the circumstance of numerous relics having been found in the immediate vicinity of the tree, indicating the existence, at some period, of an Indian encampment It had become so much an object of interest to the

Tulip Tree.

children, that one of the first excursions which they desired to make after coming to the school was a visit to the 'Indian Tree.'"

This tree, which must have been at least 300 years of age, was destroyed in 1845. Some of the pupils, either ignorant of the consequences, or with a mischievous craving for fun, kindled a fire in the cavity, which soon shrouded its noble form in flames. The news of this catastrophe was received by the scholars with a general outburst of indignation.

One of the most beautiful and interesting trees which decorates the English landscape, is the Yew. Its tall and majestic figure, as well as its dense and fine foliage, render it an attractive object; while the advanced age to which many have been known to attain, would naturally excite in the beholder a feeling of peculiar interest.

The Elm and the Yew are the favorite trees in the Church-yard, and there appears to be considerable appropriateness in the selection; the former, with its long pendulous and weeping boughs, harmonizes with the mournful surroundings of the tomb, while the latter, with its perennial verdure, its longevity, and the extraordinary durability of its wood, is emblematic of that unfading existence which awaits the spirits of the redeemed.

Gray, in his beautiful elegy, assigns to these a very prominent place.

"Beneath those rugged elms, that yew-tree's shade,
  Where heaves the turf in many a mould'ring heap,
 Each in his narrow cell for ever laid,
  The rude forefathers of the hamlet sleep."

THE YEW.

Perhaps the oldest tree of this kind on record is the "Fontingall Yew," which stood in a church-yard in Scotland. Its age is unknown, but it is asserted that there is strong probability of its having been a flourishing tree at the commencement of the Christian era. About the year 1790, it measured 56 feet 6 inches in circumference at the base of the trunk. It has since become very much decayed, and, in 1833, the entire central part had fallen away, leaving it with apparently two trunks which form a sort of arch, "through which the funeral processions of the Highlanders would sometimes pass."*

The famous Yews of Fountain Abbey in Yorkshire are well known. "The abbey was founded in 1132, in the midst of a rough piece of wood-land, in which grew seven large Yew-trees. In 1658, these trees were said to be of extraordinary size, the trunk of one of them being 26 feet 6 inches in circumference. At that time but six were standing, the largest having been blown down, and they grew so closely together as to form with their boughs a cover almost equal to a thatched roof. Under this shelter tradition tells us the monks resided until they had built the monastery."*

"The Ankerwyke Yew, near Stains, is supposed to be upwards of 1000 years old. Henry VIII. is

The Yew.

---

* Loudon's Arboretum.

said to have made it his place of meeting with Annie Boleyn, while she was living at Staines; and Magna Charta was signed within sight of it, on the island in the Thames between Runnymede and Ankerwyke. The girth of this tree, at 3 feet from the ground, is 27 feet 8 inches.

In the eastern part of the United States, the Yew is barely more than a small bush, seldom above a few feet high; while in the west it becomes a fine large tree of some 40 to 60 feet in height, and about 2 or 3 feet in diameter. It is here one of the most conspicuous trees of the forest. The Indians of Oregon use the wood of this tree for making bows, it being very tough, heavy, and elastic.

Many of the trees of our American forests at some seasons of the year are laden with the most beautiful blossoms, which are often very conspicuous, and sometimes diffusing a rich fragrance around them. The appearance of an apple or peach-orchard in the spring is an object so familiar, that its beauty is not appreciated by many.

The Buckeye, a species of Horse Chestnut which grows in Ohio, whose early blossoms are the resort of the Humming-birds upon their arrival from the South; the Catalpa, a familiar ornament around our farm-houses; the Kentucky Coffee, a native of the Western States; and the Pride of India, one of the most lovely objects that adorns the gardens of the South, are all showy and ornamental trees.

The Locust also is a very valuable addition to the list of our botanical friends; and its long bunches of

THE MAGNOLIA.

fragrant flowers, which hang so thickly from among its fine cut foliage, will always entitle it to our admiration.

But perhaps we may consider the Magnolia as standing at the head of the list of flowering trees, for the elegance as well as the great size of its blossoms. One species called the Umbrella Tree, produces flowers in considerable abundance, each measuring about 18 inches in diameter when fully expanded; they are pure white, and possessed of a very fine odor. Another species, which grows in the Southern States, forms a handsome tree of about 30 or 40 feet in height, and in the early summer months is loaded with its large white blossoms, which are about 5 inches in diameter. The fragrance of these flowers is such as to be quite perceptible at some distance. This tree remains green during the winter, and only drops its leaves as a new set are produced to replace them.

CHAPTER IX.

CONE-BEARING TREES — THEIR PECULIARITIES — GIGANTIC TREES OF CALIFORNIA — LOCALITIES MOST FAVORABLE TO THE GROWTH OF EVERGREENS — WHITE PINE — YELLOW PINE—LONG-LEAVED PINE—BLACK, WHITE, AND HEMLOCK SPRUCE — SILVER FIR — LARCH — CYPRESS — DURABILITY OF CYPRESS WOOD — THE CEDAR OF LEBANON.

We have spoken in a previous chapter of the effect produced upon a landscape by the variety observable in the different trees, both as to their outline and the character of their foliage. We will now notice a few of a class which perhaps, above all we have mentioned, exert a great influence in beautifying the face of the earth. They form by themselves a separate group or family known to botanists as the Coniferæ, or Cone-bearing trees, and their peculiar appearance will at once distinguish them from others. They are mostly evergreens, and their foliage consists of long, narrow cylindrical leaflets, thickly scattered around the stem, as in the various species of Pine, or of short, flat, and prickly appendages, arranged in a double row, one on each side of the stem, as may be seen in the Fir and Hemlock; or sometimes they are placed in tufts at intervals of one or two inches apart, as in the Larch, &c. Often 2, 3, 4, or 5 of these leaves are clustered together in a bunch, and wrapped around at the base with a sheath. With the fruit of these trees most persons are familiar. Some of the cones are particularly beautiful, especially those of the Cedar of Lebanon and the Norway Fir.

There are perhaps few trees which attain to more gigantic proportions than some of the varieties belonging to this class. The measurements of some recently discovered in California would be considered almost fabulous, were not the accounts substantiated by the most undoubted evidence.

A specimen of the Gigantic Wellingtonia, which was recently felled, measured about 300 feet in length, and 60 feet in circumference near the base; and the following extract of a letter, received from Dr. Winslow of California, gives dimensions still more extraordinary. "There are more than a hundred of these trees which may be considered as having reached the extreme limits of growth which the species can attain. One of our countrymen measured one, of which the trunk immediately above the root was 94 feet in circumference. Another which had fallen from old age, or had been uprooted by a tempest, was lying near it, of which the length from the roots to the top of the branches was 450 feet. A great portion of this monster still exists, and, according to Dr. Lapham, the proprietor of the locality, at 350 feet from the roots the trunk measured 10 feet in diameter.

GIGANTIC TREES OF CALIFORNIA.

Wellingtonia.

By its fall, this tree has overthrown another not less colossal, since, at the origin of the roots, it is 40 feet in diameter. This, which appears to me one of the greatest wonders of the forest, and compared with which man is but an imperceptible pigmy, has been hollowed by fire throughout a considerable portion of its length, so as to form an immense wooden tube of a single piece. Its size may be imagined when it is known that one of my companions, two years ago,

GIGANTIC TREES OF CALIFORNIA. 115

rode on horseback in the interior of this tree for a distance of 200 feet, without any inconvenience. My companions and myself have frequently entered this tunnel, and progressed some 60 paces, but have been arrested before reaching the end by means of wood which had fallen from the ceiling. Near these overthrown giants others still are standing not inferior to them in size, and of which the height astonishes the beholder."

In reading of a tree 450 feet high, and 30 feet in diameter, we are struck with large figures, but we must have some familiar object with which to compare it before we can fully realize the magnitude of such an object. If the tree above described was placed in the chasm of the Niagara river at the Suspension Bridge, it would stand 200 feet above the top of the bridge. By the side of Trinity Steeple in Broadway, New York, it would overtop it by 160 feet. It would be 230 feet higher than Bunker Hill Monument, and 270 feet above the Washington Monument in Baltimore. If cut up for fuel, it would make at least 2500 cords, or as much as would be yielded by 60 acres of good woodland. If sawed into inch boards, it would yield about 2,500,000 feet, and furnish sufficient 3-inch plank for 25 miles of Plankroad. This is quite enough for the product of one little seed less in size than a grain of wheat.

By counting the annual rings, it appears that some of the oldest specimens have attained the age of 3000 years. If this computation be correct, and there is probably no good reason to doubt it, they must cer-

116 GIGANTIC PINE—WHITE PINE.

tainly have existed in the days of the Prophet Elijah, or even as Dr. Lindley observes of the tree *first* described, "It must have been a little plant when Samson was slaying his Philistines."

Gigantic Pine.

On the Pacific coast there is found a species of Pine, very similar in its appearance to our common White Pine, which grows to the height of 200 feet. The trunk of a specimen which had been overturned by the winds measured 215 feet in length, and 57 feet in circumference at 3 feet from the base, and at 134 feet from the ground was about 6 feet in diameter. These are straight and beautifully tapering, and sometimes 170 feet without a branch. The cones measure about 16 inches in length.

White Pine, leaves arranged in fives.

One of the most pleasing characteristics of the Conifera is their evergreen foliage. When dark winter spreads a sombre veil over the landscape, how charming and enlivening is the effect produced by a few Pines and Firs! In cold climates, where the winters are long, and the ground is covered mostly with snow-drifts, the

THE YELLOW PINE.

general dreariness of the aspect is relieved by the abundance of the evergreens. This appears to be a special provision of Providence to give additional comfort to the inhabitants, as it is observed that these trees delight in cold and elevated positions.

In the mountainous districts of the Northern and Eastern States; upon the Alleghanies, the Rocky Mountains, the Sierra Nevada of California, and the Table Lands of Mexico; on the lofty Himalaya Mountains, the snow-crowned hills of Norway, and on the far-famed heights of Mount Lebanon, may be seen in the greatest perfection some of their most beautiful and wonderful forms.

Of those which are most familiar to us as natives of our own land, are the White, Yellow, and Long-leaved Pines; the White, Black, and Hemlock Spruce; the Silver Fir, the Larch, and the Cypress; (the two latter are not evergreen). Besides which, there are many others of less value and importance. Upon the White and Yellow Pine

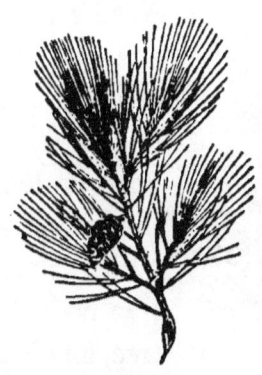

Yellow Pine, leaves arranged in twos.

we are dependent for a large amount of the lumber used in building both our houses and ships. The Hemlock also is a valuable tree to the ship-builder, as its trunk is remarkably straight, and gradually tapering toward the summit, which makes it particularly serviceable for masts and spars.

Silver Fir.

From the Yellow and Long-leaved Pines we have our supply of turpentine; and from the Silver Fir is obtained a very useful resinous substance, known as Balm of Gilead, or Canada Balsam.

The Larch and the Cypress, although cone-bearing trees, differ from the rest in being what are called "deciduous," which means that they lose their foliage every year.

In the summer season, the Larch is one of the most beautiful trees that graces the forests of the Northern and Eastern States. Its tall straight shaft, sometimes 100 feet high, and 3 feet in diameter at the base, with its minute foliage, which is densely arranged upon its long and slender branches, together with the perfect symmetry of outline which it often assumes, render it an attractive object. It is a comparatively rare tree, and is not known to exist much south of the latitude of Philadelphia, except where it has been planted as an ornament.

Larch.

In the warmer parts of the United States, the place of the Larch is supplied by the Cypress. In South Carolina, Georgia, Florida and Louisiana, this tree grows in immense quantities in the low swampy grounds contiguous to the large rivers. These "Cypress Swamps," as they are called, often occupy thousands of acres. In this rich alluvial soil, upon which a new layer of vegetable mould is every year deposited by the floods, the Cypress attains its utmost development; the largest being about 120 feet high, and 30 to 40 feet in circumference.

The European Cypress is an evergreen : its foliage bears a close resemblance to our common Red Cedar. It grows in various parts of the country adjacent to the Mediterranean Sea. It is most abundant on the islands of Crete and Cyprus, from the latter of which it derives its name. It is also spoken of in Ecclesiasticus as growing on Mount Sion. The Gopher Wood of which Noah built the ark is supposed to be identical with the Cypress. The great durability of the wood rendered it peculiarly serviceable to the ancients. Pliny, the Roman historian, says that the statue of Jupiter, in the Capitol at Rome, which was of Cypress, had existed above 600 years without exhibiting any signs of decay. Plato, a heathen philosopher, had his laws engraven on Cypress-wood as being more durable than brass. Leon Alberti, a celebrated Florentine architect of the fifteenth century, tells us that he found the wood of a vessel which had been submerged 1300 years, and which was perfectly sound, to be principally of Cypress. The Cypress doors of St. Peter's Cathedral at Rome, which were removed by Eugene IV., after having stood the usage of over 1100 years, were entirely sound ; and it was the custom in the middle ages to bury the Popes in coffins of Cypress, under the belief that they would never decay.

The Cypress is often a long-lived tree, although it will sometimes attain a great size in a comparatively short time. A tree of the American variety, planted by John Bartram, in his botanic garden near Philadelphia, some 100 years since, now (1859) measures about 9 feet in diameter, and over 100 feet in height. An old and venerated tree of the European species was some years since standing near Somma, in Lombardy, which was supposed to have been planted the year of the birth of our Saviour, although it is said that a record exists at Milan which proves that it was a tree in the time of Julius Cæsar, B. C. 42. So great was the respect shown for this tree, that Napoleon Bonaparte, when laying down the plan for his great road over the Simplon, diverged from the straight line to avoid injuring it.

In the Scriptures we find frequent allusions made to the Pine, the Fir, the Cypress, and the Cedar, all of which appear to be natives of Syria. In Isaiah xli. 19, the Pine, the Fir, and the Cedar are spoken of; and again in lx. 13, "The glory of Lebanon shall come unto thee, the fir tree, the pine tree, and the box together, to beautify the place of my sanctuary." But of all the trees of this class, the Cedar of Lebanon seems to have been regarded by the Sacred writers as a tree of uncommon beauty, and was therefore frequently used in the figurative language of the times to convey the idea of majesty and power.

In Ezekiel, chap. xxxi., we have the following remarkable expressions: "Behold the Assyrian was a Cedar in Lebanon, with fair branches and with a shadowing shroud, and of an high stature, and his top was among the thick boughs: his height was exalted above all the trees of the field, and his boughs were multiplied and his branches became long, because of the multitude of waters when he shot forth. The Cedars in the garden of God could not hide him: the Fir trees were not like his boughs, and the Chestnut trees were not like his branches; nor any tree in the garden of God was like unto him in his beauty." In this beautiful description, two of the principal characteristics of the Cedar of Lebanon are marked; viz., the length and number of its branches, and the wide expanse and density of its shade. Few trees spread themselves so thickly upon every side. This is in consequence of the horizontal growth of the branches, which shoot out in great numbers from the parent stem, forming a deep and quite impenetrable shade. These branches sometimes droop so as almost to reach the ground.

Cedar of Lebanon.

It is supposed that some of the trees still standing on Mount Lebanon are the remains of the forests from which Solomon obtained the wood for the building of the Temple. These are protected with great care, and are accounted sacred by the inhabitants. But they are gradually diminishing in numbers, and almost every few years witnesses the removal of one or more of these interesting relics, which yield to decay that strength which has defied the blasts of ages.

Of those whose appearance warrants the belief that they are the *very* Cedars under whose shade the Patriarchs of old have rested, in 1550 there remained about 28, in 1745 there were but 15. Twelve were recently counted by a traveller (Lord Lindsay), who, speaking of them, remarks, that he and his companions halted under one of the largest of them, inscribed on one side with the name of Lamartine. The grove was composed of trees of various ages growing together; "One of them," he says, "by no means the largest, measured $19\frac{1}{4}$ feet in circumference, and in repeated instances, two, three, or four large trunks spring from a single root. Of the giants there are several standing very near each other, all on the same hill; three more a little further on, nearly in a line with them; and in a second walk of discovery, I had the pleasure of detecting two others, low down on the northern edge of the grove. Lamartine's tree is 49 feet in circumference; and the largest of my two on the southern slope is 63 feet, following the irregularities of the bark."

This Cedar grows not only on the mountains of Lebanon, but also on Mounts Amanus and Taurus, in Asia Minor, in some parts of Africa, and on the islands of Cyprus and Crete. It loves cold and mountainous places, and on Mount Lebanon it grows freely among the snow.

With some of the inhabitants of the East the wood of this tree has the reputation of never decaying, and it certainly must be possessed of great durability, or it would not have been selected by Solomon for the many purposes in which he used it, where that property seemed requisite. It is described as soft and fine-grained, and sometimes beautifully marked with waving lines. It has an agreeable smell, and indeed everything about the trees has a strong balsamic odor, giving a delightful fragrance to the air in their vicinity. This is most probably the smell of Lebanon spoken of in Cant. iv. 11, and Hosea xiv. 6.

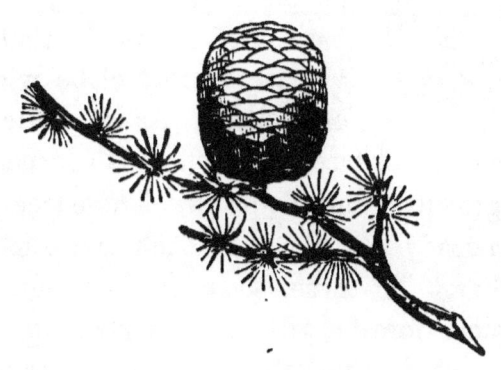

Cone of Cedar of Lebanon.

CHAPTER X.

THE PALM TREE.

MANNER OF GROWTH — LEAVES — FRUIT — THE USES OF THE PALM—COCOANUT TREE—DATE TREE—SAGO PALM—CANES FOR CHAIR BOTTOMS—DRAGON'S BLOOD—THE FIXTURES OF AN INDIAN COTTAGE—LOCALITIES OF THE PALM—THE PALMETTO—NUMBER OF SPECIES OF PALMS—GENERAL CHARACTERISTICS—THE PALM TREE OF THE BIBLE.

THE interest which we take in the study of Nature will be found to be much increased, by the comparison of the productions of one quarter of the globe with those of another. As in smaller plants, so also in trees, each clime is adorned with a growth peculiarly its own. Those trees we have already described are such as are most familiar. To these, some of the productions of the tropical forests will form a striking contrast, among which is the Palm, whose lofty summit rears itself far above its surroundings, presenting the appearance, as Humboldt observes, of one forest above another. We can form but little idea of the beauty of this stately tree, or of the multiplicity of forms which it assumes, while they all partake of the same general outline and character.

It will be remembered that in speaking of the growth of trees, we mentioned that the fresh deposits of wood are made on the external surface of the trunk, immediately under the bark. This is the case with all the trees, with a few exceptions, found in the Temperate Zone, and they belong to the Exogenous plants. But the Palm may be regarded as the type of the third natural order, called Endogenous, which has been heretofore described; and by reference to the adjoining cut, the peculiar arrangement of the particles of the wood may be contrasted with that of the exogenous tree. In the exogenous, the centre or heart-wood is the hardest; the new growth is more spongy, while the bark is quite soft. In the endogenous, the exterior is hard and tough, and the interior is soft, and often pithy.

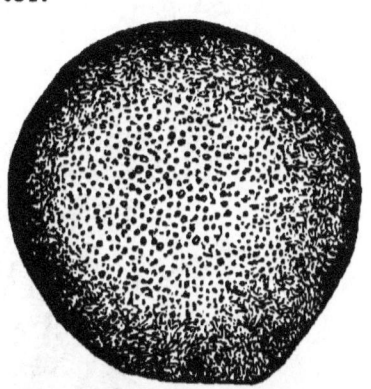

Endogenous Wood.

The Palms are lofty and erect trees, with slender, cylindrical stems, simple or rarely forked, and are marked more or less distinctly with knots or rings, which show the position of leaves which have decayed and fallen, and also indicate the progress of their growth. The leaves, which are large, often

Tupati Palm—Leaves fifty feet in length.

gigantic, sometimes measuring 50 feet long and 8 wide, are of various shapes; the largest are pinnate, or divided like a fern into long narrow leaflets; others are entire, and measure 30 feet long, and 4 to 5 wide. The pinnate leaves often assume the form of an immense fan, which, from their size and apparent lightness, are objects of great beauty.

The leaves appear in huge bunches or tufts at the summit of the tree, and are seated on long petioles or footstalks, and a constant succession is produced from the centre of this tuft, to supply the place of the lower ones, as they decay and fall off.

The flowers, which singly are small and inconspicuous, generally appear in such dense clusters as to render them peculiarly striking, especially when newly opened, as they often emit a very powerful odor.

The Palm trees are as essential to the comfort of the inhabitants of the countries in which they grow, as our most useful trees are to us. To the Wandering Arab they afford both food and shelter; and, as he sits beneath its refreshing shade, and listens to the tales of luxury and ease enjoyed by the nations of the *civilized* world, when he is told that they have no Date trees, he turns with a contented heart to the barren sands of his own dear desert, probably wondering how they endure the privation.

Perhaps we cannot better illustrate our subject than by quoting a few pages from an interesting work entitled, "The Palm Trees of the Amazon, and their Uses," by Alfred Russel Wallace, who, while on a recent tour of discovery to the tropical parts of South America, was so struck with the beauty and grandeur of these noble trees, that he was induced to make many close and interesting observations on their habits and uses.

"The materials for this work were collected during my travels on the Amazon and its tributaries, from 1848 to 1852. Though principally occupied with the varied and interesting animal productions of the country, I yet found time to examine and admire the wonders of vegetable life which everywhere abounded. Huge trees with buttressed stems, tangled climbers of fantastic forms, and strange parasitical plants everywhere meet the admiring gaze of the naturalist fresh from the meadows and heaths of Europe. Everywhere, too, rise the graceful Palms, true denizens of the tropics, of which they are the most striking and characteristic feature. In the districts which I visited they were abundant, and I soon became interested in them.

The Cocoa-nut Palm.

"The purposes to which the different parts of Palms are applied are very various, the fruit, the leaves, and the stem, all having many uses in the different species. Some of them produce valuable articles of export to our own and other countries; but they are of far more value to the natives of the districts where they grow, in many cases furnishing the most important necessaries for existence.

"The Cocoa-nut is known to us only as an agreeable fruit, and its fibrous husks supply us with matting, coir ropes, and stuffing for mattresses; but in its native country it serves a hundred purposes; food, and drink, and oil are obtained from its fruit; hats and baskets are made of its fibre, huts are covered with its leaves, and its leaf-stalks are applied to a variety of uses. To us the Date is but an agreeable fruit, but to the Arab it is the very staff of life; men and camels almost live upon it, and on the abundance of the date harvest depends the wealth and almost the existence of many desert tribes. It is truly indigenous to those inhospitable wastes of burning sands, which without it would be uninhabitable by man.

"A Palm tree of Africa, gives us oil and candles. It inhabits those parts of the country where the slave-trade is carried on, and it is thought by persons best acquainted with the subject, that the extension of the trade in palm oil will be the most effectual check to that inhuman traffic; so that a Palm tree may be the means of spreading the blessings of civilization and humanity among the persecuted negro race.

"Sago is another product of a Palm, which is of comparatively little importance to us, but in the East supplies the daily food of thousands. In many parts of the Indian Archipelago, it forms almost the entire subsistence of the people, taking the place of rice in Asia, corn in Europe, and maize and mandioca in America, and is worthy to be classed with these the most precious gifts of Nature to mankind. Unlike them, however, it is neither seed nor root, but is the wood itself, the pithy centre of the stem, requiring scarcely any preparation to fit it for food; and it is so abundant that a single tree often yields six hundred pounds weight.

"The canes used for chair-bottoms and various other purposes, are the stems of a species of *calamus*, slender palms which abound in the East Indian jungles, climbing over other trees and bushes by the help of the long-hooked spines with which their leaves are armed. They sometimes reach the enormous length of 600 or even 1000 feet, and as 4,000,000 of them are imported into this country (England) annually, a great number of persons must find employment in cutting them.

"Among the most singular products of palm trees are the resins and wax produced by some species. The fruits of a species of *calamus* of the Eastern Archipelago are covered with a resinous substance of a red color, which in common with a similar product from some other trees, is the Dragon's blood of commerce, and is used as a pigment, for varnish, and in the manufacture of tooth-powder. A lofty palm, growing in the Andes of Bogotá, produces a resinous wax which is secreted in its stem, and used by the inhabitants of the country for making candles, and for other purposes.

"The leaves of palms, however, are applied to the greatest variety of uses; thatch for houses, umbrellas, hats, baskets, and cordage in countless varieties are made from them, and every tropical country possesses some species adapted to these varied purposes, which in temperate zones are generally supplied by a very different class of plants. The Chip, or Brazilian-grass hats, so cheap in this country, are made from the leaves of a palm tree which grows in Cuba, whence they are imported for the purpose.

"The papyrus of the ancient Egyptians, and the metallic plates on which other nations wrote, were not used in India, but their place was supplied by the leaves of palms, on whose hard and glossy surface the characters were inscribed with a metallic point. These leaves, when strung together, form the volumes of a Hindoo library.

"We have now glanced at a few of the most important uses to which Palms are applied, but in order to be able to appreciate how much the native tribes of the countries where they most abound, are dependent on this noble family of plants, and how they take part in some form or other in almost every action of the Indian's life, we must enter into his hut and inquire into the origin and structure of the various articles we shall see around us.

"Suppose then we visit an Indian cottage on the banks of the Rio Negro, a great tributary of the river Amazon, in South America. The main supports of the building are trunks of some forest tree of heavy and durable wood, but the light rafters over head are formed by the straight cylindrical and uniform stems of the Jará palm. The roof is thatched with large triangular leaves, neatly arranged in regular alternate rows, and bound to the rafters with sipós or forest creepers; the leaves are those of the Caraná palm. The door of the house is a framework of thin hard strips of wood neatly thatched over; it is made of the split stems of the Pashiúba palm. In one corner stands a heavy harpoon for catching the cow-fish; it is formed of the black wood of the Pashiúba. By its side is a blow-pipe ten or twelve feet long, and a little quiver full of small poisoned arrows hangs up near it; with these the Indian procures birds for food, or for their gay feathers, or even brings down the wild hog or the tapir, and it is from the stems and spines of two species of Palms that they are made. His great bassoon-like musical instruments are made of palm stems; the cloth in which he wraps his most valued feather ornaments is a fibrous palm spathe; and the rude chest in which he keeps his treasures is woven from palm leaves. His hammock, his bow-string and his fishing-line, are from the fibres of leaves which he obtains from different palm trees, according to the qualities he requires in them, — the hammock from the Mirití, and the bow-string and fishing-line from the Tucúm. The comb which he wears on his head is ingeniously constructed of the hard bark of a palm, and he makes fish-hooks, of the spines, or uses them to puncture on his skin the peculiar markings of his tribe. His children are eating the agreeable red and yellow fruit of the Pupunha or peach palm, and from that of the Assaí he has prepared a favorite drink, which he offers you to taste. That carefully-suspended gourd contains oil, which he has extracted from the fruit of another species; and that long elastic plaited cylinder, used for squeezing dry the mandiocca pulp to make his bread, is made of the bark of one of the singular climbing palms, which alone can resist for a considerable time the action of the poisonous juice. In each of these cases a species is selected better adapted than the rest for the peculiar purpose to which it is applied, and often having several different uses which no other plant can serve as well; so that some little idea may be formed of how important to the South American Indian must be these noble trees, which supply so many daily wants, giving him his house, his food, and his weapons."

The Palms may be said to be almost exclusively a tropical production, a few only being found either to the north or to the south of their limits. A beautiful species, the "Palmetto," grows in considerable abundance in South Carolina and Florida: this appears to be the only one which exists so far north on this continent.

The Palmetto.

The whole number of species yet known is about 600, of which 275 are natives of America.

The Palms present in their varied forms some of

the most graceful and picturesque, and certainly some of the most majestic objects to be found in the vegetable world. They stand out with their light, airy, and sometimes plume-like foliage, in harmonious contrast with the deep, dark, and rank growth of the underwood. Notwithstanding there is much similarity in their general character, yet the difference is frequently great. Some species attain the enormous stature of 200 feet, while others have no stems visible above ground, and display nothing but a widespreading bunch of huge leaves; some are like reeds and are no thicker than a quill, others attain a diameter of 3 feet. The trunks of some are smooth, and some are rough with concentric rings, " or clothed with a woven or hairy fibrous covering." From the trunks of other species project cylindrical spines 8 or 10 inches in length and quite sharp, which it may be supposed often interrupt the progress of the traveller, as well as prove dangerous enemies in the dark.

The bold and erect posture of the Palm tree is proverbially emblematic of perfect uprightness. Thus David says, " The righteous shall flourish like the Palm tree."

The branches of the Palm, or rather their long leaves, were also considered as emblems of victory, and were often used as such on occasions of public rejoicing. When our Saviour made his triumphant entry into Jerusalem, some of the people " took branches of Palm trees, and strewed them in the way." And in the vision of St. John, the multitude which no man could number, were seen standing before the throne, clothed with white robes, and had palms in their hands.

In the many places in Scripture where the Palm is mentioned, it undoubtedly alludes to the Date tree which was formerly abundant in Palestine, and still is a tree of frequent occurrence throughout Asia Minor, Arabia, and Egypt.

Cocoa-nuts.

INDEX.

Acorn, wonders of, 88, 89.
Acrogenous plants, 70.
African Palm-tree, 129.
Air-plants, 42.
Algæ, 70.
Aloe, 37.
    blooms but once, 37.
    cultivated in tropical America, 38.
    fibres of, used in making rope bridges, 39.
    paper made from, by the ancient Mexicans, 39.
    various uses of, 39.
Alpine plants, 73.
Amaryllis, Yellow, 55.
Amazon, Palm-trees of the, 127–133.
American Daisy, 48.
Ankerwyke Yew, 109.
Annuals, 83.
Anther, 28.
Aquarium, 60–63.
    plants best adapted to, 65.
Aquatic plants render water pure, 60.
Arctic Circle, flowers of, 72.
Arctic regions, Willow of, 102.
Aspen, 106.

Barley, 16.
Barren Pine, 78.
    how supplied with moisture, 79.
Barton on the Passion-Flower, 41.
Bashan, Oaks of, 93.
Beech, 98, 99.
Bible, origin of the term, 20.
Biennials, 83.
Bignonia, 43.
Birch, 104.
Birch-bark canoes, 105.

Black Oak, 93.
Black Walnut, 99, 100.
Bloom, how to prolong, 33.
Blossoming trees, 110.
Boddington Oak, 95.
Box-wood, 87.
Buckeye, 110.
Bud contains the entire plant in miniature, 84.
Bulbous roots, 84.
Bulrush of the Bible, 19.
Butterfly Orchis, 43.

Cabinet-ware woods, 87.
California, large trees of, 113–116.
Calla, 55.
Calyx, 27.
Canada Balsam, 118.
Canes for chair-bottoms, 130.
Canoe Birch, 104, 105.
Carolina rice-fields, 19.
Catalpa, 110.
Catkins, 27.
Cauline leaves, 35.
Cedar of Lebanon, 120–122.
    spoken of by prophet Ezekiel, 120.
Cedars of Lebanon, 121.
    number of, now remaining, 122.
    properties of the wood, 123.
Century Plant, 37.
Cereal grasses rapid in growth, 15.
Cereus, Night-blooming, 30, 31.
Chandos Oak, 95.
Charter Oak, 93.
Chestnut, 98.
    its resemblance to the Oak, 97, 98.
China Aster, 48.
Circulating fluid of plants, 58.

INDEX.

Cocoanut, 129.
Compound flowers, 49.
Cone-bearing trees, 112.
   enlivening effect of their foliage in winter, 116.
Conifera, 112.
Convolvulus, 44.
Cork Oak, 92.
Corolla, 27.
Corymb, 29.
   figured, 31.
Cowthorpe Oak, 97.
Crawley Elm, 104.
Crocus, 81.
Curled Willow, 103.
Cyme, 29.
   figured, 32.
Cypress, 118.
   a long-lived tree, 120.
   swamps of the South, 119.
Cypress-wood, durability of, 119.
   supposed to be the gopher-wood of Scripture, 119.

Daisy, 47, 48.
   a compound flower, 48, 49.
   Michaelmas, 48.
Date, 129.
Date-tree, the palm of Scripture, 135.
Dog-wood, flowers of, 30.

Egyptian Water-Lily, 54.
   venerated by the Hindoos, 54.
Elm, 103, 104, 108.
Endogenous plants, 13.
   trees, 125.
Epiphytes, 43.
European Cypress, 119.
Evening Primrose, 31.
Evergreens, 75, 116.
Exogenous trees, 86.

Fascicle, 29.
   figured, 30.
Ferns, 70, 71.
   seed-vessels of, 71.
Fibrous-rooted plants, use of, 84.
Filament, 28.
Fir, 113, 117, 120.
Flowers, 25, 26.
   abound in mountainous places, 73.
   common form of, 28.
   different arrangement of, 29.
   of Arctic countries, 73.

Flowers of the grasses, 13.
   parts of, described, 27, 28.
   seed-producing organs of, 27.
   various shapes of, 28.
Flushing Oaks, 93.
Fly-trap, 82.
Fontingall Yew, 109.
Food-plants, their power of reproduction, 14.
Fronds, 71.
Fruits, 34.

Garden 8000 feet above the sea, 74.
Grass, meaning of the term as used in Scripture, 12.
   many varieties of, 12.
Grasses, abundance of seeds, 14.
   formation of, 13.
   order of, in creation, 11.
Great Water-Lily, 51.
Gopher-wood, 119.

Hemlock, 117.
Hickory, 100.
Himalaya Mountains, rhododendrons of, 75, 76.
Hindoo paper, 131.
Hollow Elm at Hampstead, 104.
Holly-leaved Oak, 92.

Indian Corn, 17.
Insect-catching plants, 82.
Involucre, 30.
   figured, 34.
Ipomæa, 45.

Jacobean Lily, 56.
   curious process of fertilization, 56.
Jará Palm, 131.

Larch, 118.
Large trees of California, 113–116.
Leaves, breathing-organs of plants, 58.
   of endogenous plants, 13.
   of the fir, 112.
   of the larch, 113.
   of the palm, uses of, 130, 131.
   of the pine, 112.
   of various species of oak, figured, 88–93.
   varieties of, 35.
   various shapes of, 35, 36.
Lebanon, cedars of, 121–123.
Lichens, 70.

INDEX.

Light not essential to all flowers, 30.
Lily, the emblem of purity, 50.
    many varieties, 51.
    of the New Testament, 55.
    of the Old Testament, 54.
Live Oak, 91, 92.
Locust, 110.
Lombardy Poplar, 103, 106.
Lotus, 54.
Loudon's Arboretum, quoted, 94–97.

Magdalen Oak, 96.
Magnolia, 111.
Maize, 17.
Marine plants, 65.
Marvel of Peru, 32.
Meadow-grass, recuperative power of, 23.
Merton Oak, 94.
Michaelmas Daisy, 48.
Milfoil, 66.
Mimosa, 82.
Morning-Glories, 44.
Mosses, 70.
Mould, 70.
Mount Etna, famous chestnut-tree of, 98.
Mummy wheat, 15.
Myriophyllum, 66.

Napoleon's Willow, 102.
Nature's power of adaptation, 78, 79.
Nectary, 27.
Night-blooming Cereus, 30, 31.
Night-blooming flowers, 30.

Oak, growth of, described, 89, 90.
    in Syria, confined mostly to elevated positions, 92.
    its resemblance to the chestnut, 97, 98.
    varieties of, 88.
Oaks, several celebrated English, described, 94–97.
Oats, 17, 18.
Old trees, 94, 96, 98, 99, 108, 109, 120, 122.
Orchids, 42, 43.
Orders of the vegetable kingdom, 12.
Ornamental trees of American forests, 110.
Ovary, 27.
Oxygenation of air by leaves, 60.

Palestine, oaks of, 92.

Palm, 124–135.
    almost exclusively tropical, 133.
    emblem of uprightness, 134.
    leaves of, 126.
    of Scripture, 135.
    utility of, 127.
Palm-branches emblems of victory, 134.
Palms, leaves of, formerly used in India as paper, 131.
    of South America, 131, 132.
Palmetto, 133.
Pampas Grass, 22.
Panicle, 29.
    figured, 32.
Paper Birch, 104.
Paper of the ancients, 20.
Papyrus, 19–21.
Pashiúba Palm, 132.
Passion-Flower, 40, 41.
    Barton's verses on, 41.
Peduncle, 29.
Perennials, 83.
Perfumes highly esteemed by the ancient Jews, 81.
Period of rest in plants, how indicated, 74.
Petals, 27.
Picture-writings of the Mexicans, 39.
Pines of the United States Pacific coast, 116.
Pistils, 27.
Pitcher Plant, 79.
Plants adapted to air and water, 66.
    evolve oxygen in daytime, 59.
    of cold countries usually have white flowers, 73.
    power of motion in, 81–83.
    resemblance of their functions to those of animals, 58.
    respiration of, 58, 59.
    their season of rest, 74.
Plato's laws engraved on cypress-wood, 119.
Pollen, 28.
Pond Lily, 54.
Poplar, 103, 106, 107.
    Lombardy, 103, 106.
Pride of India, 110.
Primrose, Evening, 31.
Pulque, 40.

Raceme, 29.
    figured, 30.
Radical, 35.
Reflected flowers, 28.

INDEX.

Respiration of plants, 58, 59.
Rhizome, 71.
Rhododendron, 75.
Rice, 18.
Rice-fields of Carolina, 19.
Ringent flowers, 28.
Rings in exogenous wood denote years of growth, 91.
Roots, 83, 84.
Rope bridges of Mexico, 39.
Rose, 45–47.
    cultivation of, in India, 46.
    wild, 47.
Rose-wood, 87.

Saffron, 81.
Sago, 129.
Salcey Forest Oak, 94.
Samphire, 68–70.
Sandal-wood, 87.
Sap, circulation of, 90.
    effect of light on, 90.
Satin-wood, 87.
Scriptures, allusions to cone-bearing trees in, 120.
Sea-Aquarium, 68.
Sea-Weeds, 67, 70.
Sensitive Plant, 82.
Shapes of leaves, 35, 36.
Shellbark Hickory, 100.
Silver Fir, 118.
Sleep of plants, 32.
South American Indians, usefulness of the palm-tree to, 132.
Spadix, 56.
Spike, 29.
Spikenard, 80.
Spruce, 117.
    roots of, used in making birch-bark tents, 105.
Stamens, 27, 28.
Starwort, 66.
Stems of the grasses, 14.
Style, 27.
Sun-flower, 49.
Sweet Potato, 45.

Tea, annual product of, in China, 78.
Tea-leaves, properties of, 77.
Tea-Plant, 76.
    green and black tea product of the same plant, 77.
Tents made of birch-bark, 105.

Thyrse, 29.
    figured, 33.
Tortworth, chestnut-tree at, 98.
Transformation of an insect, 63–65.
Trees add to beauty of a landscape, 103.
    contrast of, afforded by different climates, 124.
    3000 years old, 115.
    uses of, 85, 86.
Trumpet-Flower, 43.
Tuberous roots, 84.
Tulip, 55.
Tulip Poplar, 107.

Umbel, 29.
    figured, 31.
Umbrella Tree, 111.
United States, variety of plants of, 74.

Vegetable kingdom, orders of, 12.
Vegetative power of wheat, 15.
Venus Fly-trap, 82.
Victoria Regia, 51–53.
    Bridges's account of discovery of, 51, 52.

Walnut, 99.
Water Buttercup, 66.
Water-Lily, 54.
Wax from palm-trees, 130.
Weeping Willow, 101.
Wellingtonia of California, 113.
Wheat, 15.
    not found wild, 15.
White Pine, 116, 117.
White Lily, 55.
White Oak, 93.
Whorl, 35.
Wild Roses, 47.
Willow, varieties of, 102.
Winfarthing Oak, 94.
Woods used in the arts, 87.

Yellow Amaryllis, 55.
Yellow Birch, 104.
Yellow Pine, 118.
Yew, 108.
    in the United States, 110.
Yews of Fountain Abbey, 109.

Zebra-wood, 87.

www.ingramcontent.com/pod-product-compliance
Lightning Source LLC
Chambersburg PA
CBHW020100170426
43199CB00009B/342